Australian Jewel Beetles

An Introduction to the Buprestidae

**Geoff Williams, Kevin Mitchell
and Allen M. Sundholm**

CSIRO

PUBLISHING

In dedication to Charles 'Chuck' Bellamy for his life-long
contributions to our understanding of the world's Buprestidae

Author contributions: Geoff Williams project concept, text, images; Kevin Mitchell photographic equipment design and development, stacked focus images; Allen M. Sundholm images.

A catalogue record for this book is available from the National Library of Australia.

ISBN: 9781486317400 (hbk)
ISBN: 9781486317417 (epdf)
ISBN: 9781486317424 (epub)

How to cite:
Williams G, Mitchell K, Sundholm AM (2024) *Australian Jewel Beetles: An Introduction to the Buprestidae*. CSIRO Publishing, Melbourne.

Published by:

CSIRO Publishing
36 Gardiner Road, Clayton VIC 3168
Private Bag 10, Clayton South VIC 3169
Australia

Telephone: +61 3 9545 8400
Email: publishing.sales@csiro.au
Website: www.publish.csiro.au
Sign up to our email alerts: publish.csiro.au/earlyalert

Front cover: (top, left to right) *Metaxymorpha grayii, Nascioides pulcher* (photos by Geoff Williams), *Temognatha flavicollis* (photo by Allen M. Sundholm), *Castiarina maculipennis* (photo by Geoff Williams), *Neocuris guerinii* (photo by Allen M. Sundholm); (bottom) *Stigmodera gratiosa* (photo by Kevin Mitchell)
Back cover: (left to right) *Melobasis cupricollis, Castiarina ocelligera, Castiarina commixta, Castiarina quadrifasciata* (photos by Kevin Mitchell)

Edited by Joy Window
Cover design by Cath Pirret
Typeset by Envisage Information Technology
Printed by Ingram Lightning Source

CSIRO Publishing publishes and distributes scientific, technical and health science books, magazines and journals from Australia to a worldwide audience and conducts these activities autonomously from the research activities of the Commonwealth Scientific and Industrial Research Organisation (CSIRO). The views expressed in this publication are those of the author(s) and do not necessarily represent those of, and should not be attributed to, the publisher or CSIRO. The copyright owner shall not be liable for technical or other errors or omissions contained herein. The reader/user accepts all risks and responsibility for losses, damages, costs and other consequences resulting directly or indirectly from using this information.

CSIRO acknowledges the Traditional Owners of the lands that we live and work on across Australia and pays its respect to Elders past and present. CSIRO recognises that Aboriginal and Torres Strait Islander peoples have made and will continue to make extraordinary contributions to all aspects of Australian life including culture, economy and science. CSIRO is committed to reconciliation and demonstrating respect for Indigenous knowledge and science. The use of Western science in this publication should not be interpreted as diminishing the knowledge of plants, animals and environment from Indigenous ecological knowledge systems.

Contents

	Acknowledgements	v
	Preface	vi
1	**Introduction**	**1**
	Fossil history	2
	Gondwanan and extra-continental associations	3
	Warning colouration, defence and predators	4
	Life histories	5
	Plant associations	6
	Buprestids as pollinators	11
	Movement of pollen	13
	Foraging constancy	13
	Colour section 1: Beetle specimens	**15**
2	**Composition, ecology and distribution of Australian genera**	**59**
	Subfamily Polycestinae	59
	Subfamily Chrysochroinae	63
	Subfamily Buprestinae	67
	Subfamily Stigmoderinae	82
	Subfamily Agrilinae	87
	Colour section 2: Live beetles	**99**
3	**Regional buprestid faunas**	**143**
	North Queensland and the Wet Tropics	143
	Central Queensland	146
	Central and western New South Wales	147
	Sydney and the Blue Mountains	149
	Barrington Tops	151
	New England Tablelands and associated montane rainforests	152
	Littoral rainforests of northern New South Wales	153
	Alpine and montane southern New South Wales and eastern Victoria	155
	North-western Victoria	157
	South Australia	159
	South-west Western Australia	162
	Polycestinae	163
	Chrysochroinae	163

Buprestinae 166

Stigmoderinae 168

Agrilinae 169

Tasmania 170

Colour section 3: Habitat 173

4   **Threats and conservation** 183

Appendix 1. List of buprestid genera recorded from Australia 185

Appendix 2. Summary of larval and adult plant relationships 187

Appendix 3. Pollen loads from Buprestidae collected in lowland subtropical rainforest and wet sclerophyll forest 192

Appendix 4. Early taxonomists and collectors: 1770–1950 193

Appendix 5. Divisions of geological time 200

Glossary 201

Bibliography 203

Index 214

Acknowledgements

We express grateful appreciation to our partners Thusnelda (GW), Elizabeth (KM) and Adelaida (AMS) and our families for their support during the years spent either in the field, or in developing the photographic expertise and specimen preparation that allowed the fulfilment of this project.

However, the scope of this book would not have been possible without the preceding studies and collecting efforts of taxonomists and field naturalists past and present, Australian and foreign. Many of the early workers are profiled in the text or in Appendix 4, but to those friends and colleagues who over the years have contributed information, provided company during long hours on the road or supplied specimens to photograph, we extend a collective acknowledgement and debt. We thank Hugh Nicholson (The Channon) and Brett and Marie Smith (Ellura Sanctuary) for providing additional location photos, Simon Grove (Tasmanian Museum and Art Gallery) for photos of *Castiarina flavopicta*, *C. insularis*, *C. insculpta*, *C. jubata*, *C. macquillani*, *C. rudis*, *C. tasmaniensis*, *C. virginea*, *Nascioides quadrinotatus*, *Synechocera deplana*, and *Temognatha mitchellii*, Geoff Thompson and Lily Kumpe (both Queensland Museum) for facilitating the images of the holotype of *Nascioides elessarellus*, Robert Richardson for the image of *Temognatha duponti*, Geoff Monteith (Queensland Museum)

for help and information with various matters buprestid-wise, and Natalie Tees (Australian Museum) who, in addition to answering numerous specimen queries, undertook the photography of *Anthaxomorphus queenslandicus*, *Castiarina rudis*, *Euryspilus chalcodes*, *E. viridis*, *Nascioides parryi*, *Paracupta aurofoveata* and an enigmatic *Anilara*-related species, and resurrected an ageing syntype of *Nascio chydaea*. In these latter tasks Russell Cox, Clare Kim and Derek Smith (Australian Museum) are thanked respectively for helping out with general specimen enquiries and for assistance with specimen preparation. Kimbiri Pullen (Research Fellow, CSIRO) kindly reviewed the section on New South Wales and eastern Victorian alpine fauna. Roman Hołyński (Graniczna, Poland) helped with queries regarding the genus *Metataenia*. CSIRO Publishing is thanked for permission to use the images of *Anthaxoschema terrareginae*, *Aphanisticus endeloides*, *Australorhipis aphanochila*, *Barakula petersonorum*, *Burnsiellus marmorata*, *Endelus* sp., *Hedwigiella jurecki*, *Microcastalia globithorax*, *Strigoptera bimaculata* and *Toxoscelus queenslandlicus*.

And far from least, as our dedication indicates, we honour the late Charles 'Chuck' Bellamy for his extensive contribution to our understanding of Australia's, and the world's, diverse buprestid fauna.

Preface

Australian Jewel Beetles: An Introduction to the Buprestidae has been a cooperative effort in which we have sought to combine our individual knowledge and experience, be this in the realms of invertebrate taxonomy and ecology, or photography. It is intended as an introduction to the interrelationships and diversity of Australia's spectacular buprestid fauna, from which we trust will follow a greater concern for the fauna's conservation, and that of the landscapes and plant communities in which buprestids dwell and interact.

However, the book is not a field guide, nor does it provide instruction on morphology. Our buprestid fauna is far too diverse, often with distinguishing characters accessible only by microscope (as examples among the included photographs demonstrate), to allow placement within a single volume suitable for field use. Nevertheless, extensive details of structure and form can be found in Lawrence and Ślipiński (2013) and Ślipiński and Lawrence (2019). The landscapes and plant communities in which jewel beetles are to be found are equally diverse, and so we have included an array of panoramas and vegetation formations (Figs 523–576) as informative, though far from exhaustive, examples of those areas that we have visited.

The Australian buprestid literature is a large one and includes numerous papers by foreign workers, those dealing with taxonomy often resulting in a multitude of synonyms. We treat the fauna systematically by subfamily (after Bellamy 2002, 2003), but with genera placed alphabetically there within. Examples of subfamilies, genera and species are given either as 'stacked focus' images (Figs 1–264) and/or live individuals (Figs 265–522). The majority of beetle images are dorsal in orientation; however, there is a subset of comparative photos in which the subject is also photographed from the side. All genera recorded from Australia are illustrated. Taxa above genera are considered as nominative plural (Cranston *et al.* 1991). Nomenclature generally follows Lawrence and Lemann (2019). A detailed illustrated key to Australian genera is provided in the same reference. A summary of adult and larval food plants has been given by Bellamy *et al.* (2013).

Recent revisions and reviews over the last several decades include, but are not limited to, those of the genera and tribes *Agrilus* (Curletti 2001), *Astraeus*, *Castiarina* and *Diphucrania* (Barker 1975, 1989, 2001, 2002, 2006a, 2006b, 2006c), *Araucariana*, Epistomentini, *Melobasis* and *Prospheres* (Levey 1978a, 1978b, 2012, 2018), *Bubastes* (Bílý and Hanlon 2020), *Calodema* and *Metaxymorpha* (Nylander 2008), *Chalcophorella* (Toyama 1986), *Nascioides* (Williams 1987), *Neobuprestis* and *Burnsiellus* (Levey and Bellamy 2013), *Australorhipis*, Chalcophorini, *Dinocephalia*, *Meliboeithon*, *Paracephala*, *Synechocera* (Bellamy 1986, 1987b, 1988), *Stanwatkinsius* (Barker and Bellamy 2001), Stigmoderini (Gardner 1989) and *Xyroscelis* (Williams and Watkins 1986; Bellamy 1997). Additional publications pertinent to the taxonomic status, distribution or ecology of the Australian fauna have been authored or co-authored by Svatopluk Bílý, David Cowie, Trevour Hawkeswood, Roman Hołyński, David Knowles, Peter Lang, Eric Matthews, Magnus Peterson, Michael Powell, James Turner, Mark Volkovitsh and Tom Weir. These are cited in the text and Bibliography.

Much of the following text concerns plant associations, for buprestids are phytophages, and as larvae and adults, they are obligate consumers of plant material. So it is valuable to remind readers that in a paper published in 2013 in *Insecta Mundi* Chuck Bellamy and his colleagues warned: 'Because of the dynamics of both animal and plant nomenclature, there are several concerns that should be stated, so users of this work don't accept it completely without recognising there are some aspects that are not so easily verified. It is a nearly impossible task to reconcile earlier host records under buprestid names that may well be wrong in light of modern revisions: e.g. *Anilara*, *Bubastes*, *Diphucrania*, *Selagis*, and *Temognatha*.' Plant names and the definition of plant families and genera have also evolved, though not always without contention or finding universal resolution. Further, there is the concern that beetle and or host plant identifications, in the first place, were just simply wrong.

Thus the cautionary tale of Bellamy *et al.* (2013) holds equally true for this book.

1   Introduction

Beetles are the largest group of insects and constitute the order Coleoptera. Worldwide there are nearly 450 000 described species (Ślipiński *et al. 2011*); however, this is only a small portion of the likely total number. Of the currently recognised 160 families of Coleoptera 117 occur in Australia (Lawrence and Ślipiński 2013).

In company with the Schizopodidae (Fig. 260), which are restricted to south-west North America and sometimes treated as a subfamily, Buprestidae constitute the superfamily Buprestoidea (e.g. see Cobos 1980; Hołyński 1988; Jendek 2001; Bellamy 2002, 2003, 2008a, 2008b, 2008c, 2008d, 2009; Lawrence and Ślipiński 2013). Commonly referred to as *jewel beetles*, Buprestidae are one of the most popular families of beetles (see Appendix 4) (Sundholm and Catford 1982; Knowles 1984). They rank as the eighth largest family of Coleoptera, comprising about 15 000 species in more than 500 genera placed in 7 subfamilies (here reinstating the Stigmoderinae). Julodinae, Galbellinae and the closely related small family Schizopodidae or 'false jewel beetles' (7 spp., 3 genera) (Nelson and Bellamy 1991; Kolibač 2000; Bellamy 2003) are absent from Australia (Bellamy 2003). But the number of recognised subfamilies worldwide has at times differed substantially – for example, 4 (3 if excluding Schizopodinae) by Hołyński (1988); 7 by Bílý (1974); 12 by Jendek (2001); 13 by Cobos (1980), 2 by Carter and Théry (1929), and 5 by Matthews (1985). The classification of subfamilies and intermediate ranks of lower hierarchy is still to be fully resolved (Evans 2017; Lawrence and Lemann 2019).

Buprestidae have a cosmopolitan distribution, being found in all of the major biogeographical regions, and are currently so widespread and speciose that one can only contemplate what vanished faunas once occupied the now dead forests and woodlands of Antarctica, and what ecological relationships they may have partnered with now-extinct plants (Stilwell and Long 2011). Although many species are spectacularly coloured and patterned (*Asamia pulcherrima*, Africa) and are quite large (*Julodimorpha saundersii*, *Temognatha heros*, Australia; *Euchroma gigantea*, South America), there are numerous minute taxa that are dark in tone and seemingly without any conspicuous visual characters of note – except to taxonomists and systematists.

In a world context most adult buprestids are variably elongate in form (typified by *Acmaeodera*, *Agrilus*, *Anthaxia*, *Buprestis* and *Chrysobothris*), though there are several that diverge from this body form – for example, the rotund species of African *Stenocera* and *Julodis*, Madagascan *Polybothris auropicta* with its oddly explanate elytra, *P. dilata* with its broad angular pronotum (see Bellamy 2003), and the small but shield-like *Pachyschelus collaris* from Mexico (Fig. 252) and *Paratrachys* from eastern Australia (Figs 13, 14). Almost universally among the family the wing covers (elytra) fully conceal the abdomen, though in several the tip of the abdomen, the pygidium, is exposed or nearly so (as in some *Neocuris, Dinocephalia, Synechocera, Maoraxia* [females] and also *Barakula*). But in the Australian *Selagis splendens* (Fig. 114) the elytra are greatly foreshortened, a degree of reduction also exhibited by the North American genus *Hesperorhipis* (Bellamy 2003). Larvae are elongate and narrow (and in some instances may be confused with those of Cerambycidae), with most being characterised by a laterally expanded and somewhat flattened thorax that is normally distinctly wider (though to varying degrees) than the parallel-sided abdomen (see Volkovitsh and Hawkeswood 1999; Bílý and Volkovitsh 2003; Volkovitsh *et al.* 2003; Bílý and Volkovitsh 2005; Lawrence and Ślipiński 2013; Volkovitsh and Bílý 2015). Leaf and small steam-mining forms, such as *Paratrachys* (Polycestinae), *Habroloma, Trachys* and *Aphanisticus* (Agrilinae), may differ by lacking the pronounced expansion of the thorax and with individual abdominal segments being somewhat laterally convex (Bílý 1986, 1989). Leaf-mining has arisen independently in the Polycestinae and Agrilinae, with the leaf-mining Agrilinae comprising diverse and highly specialised taxa possibly indicating adaptive radiations (Evans *et al.* 2015).

This level of diversity and form is equally to be found among the Australian Buprestidae with the fauna characterised by large species (up to nearly 7 cm in size) in genera such as *Julodimorpha, Calodema, Metaxymorpha, Temognatha* and *Pseudotaenia*, yet with others that are minute, such as with those of *Aphanisticus, Endelus, Germarica, Trachys* and *Habroloma* (<3 mm). Placed within these extremes of size are members of brightly coloured *Castiarina*, metallic-hued *Melobasis*, and numerous other genera of intermediate size. In elongate form buprestids superficially resemble Elateridae, Throscidae and Eucnemidae but the DNA sequencing study of McKenna *et al.* (2015) collectively places Buprestoidea (Buprestidae + Schizopodidae) as sister to Byrrhoidea (e.g. Byrrhidae, Callirhipidae, Heteroceridae, Limnichidae and Psephenidae; also see Lawrence and Ślipiński 2013), with Buprestidae having earlier, and tentatively, been classified within Byrrhoidea by Lawrence (1988). Relationships are discussed in Lawrence and Lemann (2019).

Bellamy (2002) documented just over 1200 Australian buprestid species and discussed earlier concepts of the family. Lawrence and Lemann (2019) placed the Australian fauna in four subfamilies, comprising 74 genera; their subfamily recognition of Polycestinae, Chrysochroinae, Buprestinae and Agrilinae differing from that of Bellamy

(2002) (i.e. Polycestinae, Chalcophorinae,[1] Buprestinae and Agrilinae; collectively in about 75 genera), Bellamy (1986) and Lawrence and Britton (1994) (Polycestinae, Chalcophorinae, Buprestinae, Mastogeninae, Trachyinae, Agrilinae, Chrysobothrinae), Matthews (1985) (Agrilinae, Buprestinae, Chalcophorinae, Chrysobothrinae, Polycestinae), Théry (1929) (Buprestinae and Chalcophorinae only), and the worldwide recognition by Bílý (1974) of Polycestinae, Sternocerinae, Buprestinae, Chrysobothrinae, Agrilinae and Trachyinae.

In this volume we separate Stigmoderinae (previously Stigmoderini) as distinct from Buprestinae, the latter a polyphyletic group (Evans *et al.* 2015), thus recognising five subfamilies in the Australian fauna (i.e. Agrilinae, Buprestinae, Chrysochroinae, Polycestinae and Stigmoderinae [Appendix 1]). Stigmoderinae have been separately recognised by Britton (1970) and others (e.g. Fowler 1912; Tillyard 1926; Brues *et al.* 1954), the distinction here based on the possession of a highly derived ovipositor, a unique delineating character within the family, and which Gardner (1989) considered provided evidence of monophyletic status. In addition, Volkovitsh *et al.* (2003) note larval diagnostic features that characterise Stigmoderinae 'as one of the most specialised and isolated buprestid taxa'. Separation of the Stigmoderinae also confirms the distinct Gondwanan affinity of the included Australasian and South American stigmoderine taxa. See Gardner (1989), Volkovitsh *et al.* (2003) and Hołyński (2008) for a discussion of Stigmoderinae/Stigmoderini.

FOSSIL HISTORY

Fossils of adult Buprestidae are known, for example, from the Upper Cretaceous of Kazakhastan (*Cretofrontolina kzyldzharica*) and Russia (*Metabuprestium arkagalense, M. sibiricum*), the Lower Cretaceous of Mongolia (*Metabuprestium ichbogdense, Pseudomongoligena schinkhudukense, Trapezetergum grande*), the Middle Jurassic of China (e.g. *Sinoparathyrea bimaculata*) and the Late Jurassic of Kazakhastan (*Jurabuprestis karatauensis*) (Alexeev 2000; Grimaldi and Engel 2005; Alexeev 2009; Pan *et al.* 2011; Yu *et al.* 2013). Records attributed to the Triassic are uncertain. More geologically recent fossil buprestids are recorded in Eocene oil shales from Germany, with other compression fossils known widely from the Tertiary of Europe and North America

---

1 Toyama (1987) placed Chalcophorinae in Buprestinae.

(e.g. *Ancylocheira* from the Miocene of Germany, *Chrysobothris* spp. from the Miocene of Colorado). A schizopodid impression fossil (*Mesoschizopus elegans*) has been described from the Lower Cretaceous Yixian formation of China; this record demonstrates how extant groups with highly restricted ranges (i.e. Schizopodidae – California) had ancestral distributions that were probably more widespread in the Mesozoic (Cai *et al. 2015)*. The records of Buprestidae from amber deposits are few. These include adults of *Acmaeodera, Agrilus, Anthaxia, Buprestis, Chrysobothris, Mastogenius* and *Poecilonota*, as well as the schizopodid *Electrapate* (Bellamy 1995, 1999); these records are of non-Australian species. Fossil larvae in amber are known for the buprestid genera *Buprestis* and *Phaenops* (cited in Bellamy 1995).

Australian fossils that have been attributed to Buprestidae include *Mesostigmodera typica* from New South Wales and Queensland (Etheridge and Olliff 1890; Dunstan 1923) and *Lobites granulatus, L. trivittatus* and *L. tuberculatus* from Queensland (Tillyard and Dunstan 1923). However, Gardner (1989) questioned the placement of *Mesostigmodera typica* in Buprestidae, considering that the structure of the elytra (two separate elytra alone representing the fossil evidence) more closely corresponded with *Omma* (Ommatidae–Archostemata); the Ommatidae were recently discussed by Lawrence and Escalona (2019).

GONDWANAN AND EXTRA-CONTINENTAL ASSOCIATIONS

The Australian buprestid fauna is dominated by distinct elements (Bílý and Volkovitsh 1996) that appear to have originated in or radiated from Australia, as well as a small number of genera with Oceanian, Indo-Oriental or Old World associations (Bellamy 2002). In addition to these there are a several genera (e.g. *Agrilus, Chrysobothris*) that have a cosmopolitan distribution.

With the exception of the enigmatic *Trachys blackburni* (the Papua New Guinea record given in Lawrence and Lemann 2019) no other Australian species are known to have extralimital distributions. There is a high level of generic endemism in the Australian fauna, but nevertheless several speciose 'Australian' genera also occur in nearby landmasses: *Nascioides* (New Caledonia, New Zealand), *Castiarina* (New Guinea), *Astraeus* (New Caledonia) and arguably *Melobasis* (extending to southwest Pacific, Malaysia) (Obenberger 1930; Barker 1975, 1986; Williams and Bellamy 2002; Barker 2006c;

Nylander 2006). Several genera that occur in Australia's north and along its eastern seaboard, but whose extant distribution does not assuredly attest to an Australian origin, also occur in nearby islands such as New Guinea (e.g. *Calodema, Metaxymorpha*) or New Caledonia, New Guinea, Norfolk Island (*Prospheres*) (Levey 1978a; Nylander 2008). *Maoraxia*, and possibly *Paracupta*, also fit within this group but have a wider Oceanic distribution in the south-west Pacific, including Philippines, Lord Howe Island, Fiji, Tonga and New Zealand. Of these only *Nascioides* and *Maoraxia* (the only two genera, collectively comprising two species, recorded from New Zealand) are known to occur in Australia, New Caledonia and New Zealand (Williams and Bellamy 2002); this provides additional biogeographical support for their Late Cretaceous connection (Griffiths 1974; Bellamy 1991), with Bellamy (1991) and Bellamy and Williams (1985) suggesting that the known distribution of *Maoraxia* indicates the genus is much older than the Late Cretaceous dating of the breakup of the Australian plate provided by Raven and Axelrod (1972).

The Stigmoderinae can be considered a Gondwanan or 'Southern' fauna comprising genera restricted to Australia (*Calotemognatha, Stigmodera, Temognatha*) or, with a small extension of species into New Guinea (of *Castiarina*), genera shared with eastern Australia and New Guinea and several associated islands (*Calodema, Metaxymorpha*), and a second geographically circumscribed stigmoderine group occurring in South America (*Agrilozodes, Conognatha, Hiperantha, Lasionota* [syn. *Dactylozodes*], *Semiognatha*) (see Figure examples 172–174) (Bellamy 2003). Australian Stigmoderinae, as adults, are essentially a flower-frequenting group commonly associated with Myrtaceae, but not restricted to that family. *Selagis* (Buprestinae–Curini), which is also found in association with Myrtaceae, has relatives (*Anthaxioides*) in South America.

Worldwide the Polycestinae appear to be a relict subfamily, in Australia comprising genera mostly few in number of species (e.g. *Strigoptera, Paratrachys, Polycesta, Prospheres, Xyroscelis*) but with *Astraeus* having undergone considerable recent speciation and radiation within the continent (a phenomenon also exhibited by polycestine *Acmaeodera* in North America). *Prospheres* (Prospherini, but earlier in Polycestini [Levey 1978a]) consists of four species (*P. aurantiopictus*, eastern Australia; *P. norfolkensis*, Norfolk Island; *P. chrysocomus*, New Caledonia; *P. alternecosta*, New Guinea). Levey (1978a) notes their individually distinct facies may indicate long phylogenetic separation relating to isolation of populations following the Late Cretaceous breakup of the Australian plate.

WARNING COLOURATION, DEFENCE AND PREDATORS

Many species of *Castiarina* possess what appear to be aposematic warning patterns; however, a number of species appear to participate in mimicry complexes (Batesian and Müllerian) (discussed in Williams 2020a). These have been considered as possible mimics of asilid flies (as in the *Castiarina* 'producta' species group; see Barker 2006a), lygaeid bugs of the genus *Spilostethus* (e.g. *Castiarina hypocrita* [Fig. 385]), cantharid beetles (*Castiarina mima*, *C. pertii* [Figs 149, 154] of *Chauliognathus lugubris* [Fig. 261]; and possibly *Castiarina ocelligera* [Fig. 152] of *Chauliognathus nobilitatus*), but especially within a taxonomically broader mimic complex (e.g. Coleoptera – Belidae, Cerambycidae, Meloidae, Oedemeridae, Pyrochroidae; Lepidoptera – Oecophoridae/ *Snellenia*) (see Figure examples 262–264, 521) that, in the convergent theme of red and black colours, resemble chemically protected lycid beetles of the genus *Metriorhynchus* (Fig. 262) (Nicholson 1929; Lawrence and Ślipiński 2013); for example, *Castiarina erythroptera*, *C. latipes* and *C. nasuta*. *Castiarina testacea* has been suggested as a possibly mimic of the flower buds of *Eucalyptus* spp. (F. Douglas pers. comm.). An exceptional potential mimic species is that of *Castiarina maculicollis* (Fig. 411), which resembles certain leaf-frequenting cockroaches of the genus *Ellipsidion* (Ectobiidae) (Fig. 522).

Chemical protection is also present within Buprestidae (see Brown *et al.* 1985; Moore and Brown 1985; Ryczek *et al.* 2008), this enhancing the defence offered by colouration alone. Moore and Brown (1985) found that buprestins, noxious organic compounds, were widespread in the family, even in non-mimic species, and were likely to be universal in *Castiarina*. Their study group in which these 'bitter principles' were isolated comprised *Agrilus australasiae*, *Pseudotaenia waterhousei* (as *Chalcotaenia laeta*), *Diphucrania marmorata* (as *Cisseis marmorata*), *Selagis caloptera* (as *Curis caloptera*), *Cyrioides imperialis* (as *Cyria imperialis*), *Julodimorpha bakewelli*, *Merimna atrata*, *Nascio vetusta*, *Prospheres aurantiopictus*, *Torresita cuprifera*, *Stigmodera macularia*, *Temognatha heros* (as *Stigmodera heros*), as well as 11 species of *Castiarina*.

Jewel beetles use pigments to produce colour, and reflected colour, or the perception of it, may provide defence by visual means alone. Colour reflectance can be influenced by body size (Wang *et al.* 2023); however, iridescence, common to metallic-hued Buprestidae, is not due to pigmentation, rather it is caused by structural colouration in which multi-layer cuticle reflectors (e.g. in the elytra) selectively reflect specific light frequencies in particular directions. Kjemsmo *et al.* (2020) found that iridescence may act as a dynamic disruptive form of camouflage, concealing the beetle rather than signalling its presence, so that beetles could potentially conceal themselves by choosing glossier backgrounds, such as shiny leaf surfaces with greater levels of specular reflection, upon which to rest. Thus Australian taxa, as in *Stigmodera gratiosa*, *Pseudotaenia* spp., and numerous species of *Melobasis*, may be less conspicuous to predators than their bright metallic colours suggest to humans.

The colour and pattern of some species allows individuals to cryptically 'blend in' with the vegetation upon which they rest, as is the case with *Diadoxus erythrurus* (Fig. 49) when motionless on the needle-like foliage of its *Callitris* food plant. But general form alone may serve to hide. The flattened shape of adult *Synechocera* allows them to retreat within the protective leaf sheaths of their *Gahnia* (Cyperaceae) (Fig. 540) larval and adult host plants, and although the small size or dull colouration of many species (e.g. *Anilara*, *Diphucrania*, *Germarica*, *Helferella*, *Nascio*) suggests a cryptic search image that could mitigate the risk of predation, the visual impression of several species, as with *Xyroscelis bumanna* and *X. crocata* (Figs 18–20, 280, 281), and particularly that of *Hypocisseis pilosicollis* and *H. suturalis* (Figs 247, 513, 514), resemble bird droppings in similar fashion to that exhibited by some spiders of the families Araneidae, Arkyidae and Thomisidae (Whyte and Anderson 2017).

Species may exhibit a range of active defence responses when disturbed by vibrations of the substratum upon which they rest or by visual stimuli (Crowson 1981), some simply falling from the branch upon which they rested and feigning death ('thanatosis'), or immediately taking flight – 'drop-off' strategies described by Hawkeswood (1978) as 'free fall and flight' and 'free fall and death feign'. Behavioural defence is present in at least two genera, *Astraeus* and *Dinocephalia*. Adults of these, when approached, commonly quickly topple from the sunlit branchlets upon which they rest during the day, often returning to (or close to) the same position when danger has passed. In addition, the elytra of *Astraeus* are able to spring open from their coupled position when in repose, this action flicking the adult for up to several metres (Barker 1975). Many adult *Astraeus* possess contrasting black-and-yellow-banded colouration (suggesting an aposematic adaptation) on their elytra and so are somewhat unique in being able to deploy visual, mechanical, and possibly chemical, means of predator avoidance. In addition, when handled, some buprestids will exude an odourless dark fluid from their mouth but the function and nature of this, whether it be a coordinated defence mechanism or an auto-response, is unknown.

Nevertheless jewel beetles do fall prey. Recorded predators of adults include spiders (Araenidae), assassin bugs (Reduviidae), wasps (Crabronidae), robber flies (Asilidae) and vertebrates (Lea and Gray 1936; Barker and Inns 1976; Hawkeswood 1980a; Hook and Evans 1991; Barker 2006a; Westcott *et al.* 2016; Westcott and Lavigne 2019). Hawkeswood (1980a) records an araenid spider (possibly *Araneus heroine*) feeding on *Castiarina picta*. Barker and Inns (1976) recorded the asilid fly *Phellus piliferus* feeding on a large captured Australian *Temognatha stevensii* (as *T. tibialis*); the feeding site appeared to be in the soft fold between the pro- and mesothorax, accessed by the fly keeping the elytra partly open after catching its prey. Worldwide at least 27 genera and 53 species of Asilidae are known to predate on buprestid beetles (Westcott and Lavigne 2019). In addition to that of Barker and Inns Australian asilid records include *Selagis aurifera* by *Blepharotes coriarius*, *Temognatha duponti* and *T. heros* by *Phellus piliferus*, *T. heros* and *T. murrayi* by *P. glaucus*, *T. wimmerae* by *P. olgae*, and undetermined buprestid prey by *Cerdistus margitis* and *Blepharotes* sp. (Westcott and Lavigne 2019). Lea and Gray (1936) record a specimen of *Castiarina flava* found in the gut of the meliphagid honey-eater *Myzantha flavigula*.

Volkovitsh *et al.* (2003) cite the striped possum (*Dactylopsila trivirgata*) as a specialised predator of *Metaxymorpha gloriosa* larve. Possums use their teeth to strip away bark from infested trees and hook larvae from their feeding tunnels with their elongate fourth finger. The large yellow-tailed black cockatoo (*Calyptorhynchus funereus*) is also a likely predator, with adults frequently observed in the subtropical rainforests and wet sclerophyll forests of New South Wales grasping tree trunks of understorey trees (e.g. *Glochidion ferdinandi* [Euphorbiaceae]) and then with their powerful beaks proceeding to excavate outer wood and bark until the beetle larva within is exposed (G. Williams pers. obs.).

Parasitism of Buprestidae by small wasps is also recorded, with immature stages especially vulnerable (Queiroz 2002). The wasp superfamily Stephanoidea is recorded by Grimaldi and Engel (2005) as parasitoids of Buprestidae, the superfamily known from fossils that date from the Late Cretaceous amber of New Jersey (*Archaeostephanus corae*), but whose forebears arose in the Jurassic, a date commensurate with the presence of ancestral buprestids. Specific records of microhymenoptera families as parasites include Encyrtidae, Eupelmidae. Megalyridae, and Cleonymidae/Pteromalidae (e.g. Ferrière 1947; Hadlington and Gardner 1959; Hawkeswood and Turner 1997). However, parasitoid wasps of wood-boring buprestid larvae also include larger Braconidae and Aulacidae.

LIFE HISTORIES

Adult Buprestidae are almost exclusively diurnal in behaviour, being very active during warm sunlit days. There are, however, instances of nocturnal attraction to light (*Chrysobothris* sp., *Melobasis* spp., *Pseudanilara cupripes*, *Torresita cuprifera* in Williams 1982), in particular that of *Merimna atrata* (Hawkeswood 2007a). Adults and larvae are phytophagous (though not all adults necessarily feed), with known host and food plant records being listed most recently in Bellamy *et al.* (2013), and summarised in Appendix 2. Adults can be found in association with the foliage of living plants (e.g. *Aaaaba*, *Agrilus*, *Astraeus*, *Chalcophorotaenia*, *Cyrioides*, *Diphucrania*, *Ethonion*, *Germarica*, *Habroloma*, *Helferella*, *Hypocisseis*, *Melobasis*, *Nascioides*, *Neobuprestis*, *Paratrachys*, *Strigoptera*, *Theryaxia*), dead leaves (*Anilara*, *Pseudanilara*), and fallen or standing tree trunks (*Chrysobothris*, *Nascio*, *Nascioides*, *Notographus*, *Prospheres*). These associations relate to instances of herbivory (as in *Aaaaba*, *Habroloma*), retreat to sheltered refuge positions due to daily shading or adverse climatic conditions of longer duration, or where adults are seeking out ovipositing sites (as on fallen trunks, branches). A prominent suite of taxa (e.g. species of buprestine *Neocuris*, *Selagis*, and stigmoderine *Calodema*, *Castiarina*, *Metaxymorpha*, *Stigmodera*, *Temognatha*) are to be found commonly in some seasons feeding on floral nectar and/or pollen, particularly from flowers with open readily accessible flower structures, and so may potentially serve as pollinators (see Williams 2020a, 2021). However, within the Stigmoderinae there are species with abbreviated mouthparts (that are leaf or petal chewers) and species that are nectar or pollen feeders in which the mouthparts and frontoclypeus are prognathous (i.e. 'muzzle-like', prolonged horizontally, and directed forward). These differential morphological characters are particularly evident in *Castiarina*, in which the maxillae of nectar feeders are (though sometimes variably) profusely brush-like with dense long setae, but with this condition reduced in leaf and petal chewers (Gardner 1989). The elongation of the mouthparts is also to be found in non-stigmoderine *Julodimorpha* and *Selagis*. Théry (1929) included both genera in Stigmoderini (= Stigmoderinae as here) solely on the basis of this character, but the prolongation of the mouthparts is considered convergent (Gardner 1989).

Most buprestid larvae, depending on species, mine within dead, dying and living wood (especially that of woody angiosperms), commonly feeding and excavating just below the bark but also within the heart wood. Branch diameter may strongly influence patterns of usage (Reyes-Gonzáles *et al.* 2021); however, adults rarely oviposit on wood that has been dead for some time. Freshly fallen branches and trunks with coarse bark textures that offer protection of eggs from predators such as ants are particularly favoured.

Several genera are leaf miners of vines (*Habroloma*), grasses and rushes (such as *Aphanisticus*), breed in seed stalks (*Synechocera*), with leaf mining in Buprestidae having multiple evolutionary origins (Evans *et al.* 2015; Evans 2017). Others are known to make galls (*Dinocephalia*, *Diphucrania*, *Ethonion*, *Paracephala*). Interestingly, the Indian *Habroloma myrmecophila* provides a rare instance of association with ants (termed myrmecophily) (Bílý *et al.* 2008). After mating, adult *Habroloma myrmecophila* invade damaged or newly constructed nests of *Oecophylla smaragdina* (an aggressive leaf-nesting species also occurring in northern Australia) where they lay eggs. Only adults interact with the ants, the larvae mining the leaves forming the ant nest and pupating within the mines. Although several species of *Habroloma* are recorded from Australia, little is known of their life history (Hawkeswood 2007b; Williams 2020a).

PLANT ASSOCIATIONS

There is little widespread evidence for fidelity (as opposed to short-term feeding *constancy* at a single flower resource by adult flower-frequenting species) to particular plant hosts. However, the larval and adult association of species of *Nascioides* and some *Melobasis* with Nothofagaceae, Atherospermataceae and Cunoniaceae, that of *Prospheres* and *Araucariana* with Araucariaceae, *Diadoxus* and numerous *Astraeus* with Casuarinaceae and Cupressaceae, *Cyrioides* with Proteaceae and Cunoniaceae, and *Xyroscelis* with the cycad family Zamiaceae (Morgan 1966; Levy 1978b; Williams and Watkins 1986; Williams 1987, Bellamy *et al. 2013*; 2020a; see Appendix 2) are of particular biogeographical and co-evolutionary interest.

Although several buprestid genera are associated with gymnosperms (cycads, conifers), few are recorded from cycads – a rare example being the wholly Australian tribe Xyroscelini. The Xyroscelini comprise two species of *Xyroscelis*: *X. crocata* from Western Australia and *X. bumanna* from south-eastern Australia (Williams and Watkins 1986). Both utilise *Macrozamia* (Zamiaceae) as hosts (Figs 536, 575). Cycads are an ancient gymnosperm group suggesting the association with *Xyroscelis* may be an old and co-evolutionary one. Gymnosperms arose in the Devonian (c. 400–360 mya), this being much earlier than the flowering angiosperms (Enright *et al.* 1995; Williams 2021). Although cycads suffered major extinction events during the Mesozoic and Cenozoic there was a rediversification beginning about the Late Miocene [Nagalingum *et al.* 2011]) such that extant cycads are not likely to be much older than 12 million years, with gymnosperm groups of Cenozoic age (about 70 mya to the present) significantly younger than their angiosperm counterparts (median age: 32 mya versus 50 mya [Crisp and Cook 2011]). This reduced age scenario for living cycad taxa may explain the restriction of Xyroscelini to Australia, though not dismissing an apparent co-evolutionary association.

The origin of conifers can be traced back at least to the Late Carboniferous–Permian (at least 270 mya) (Taylor *et al.* 2009). Their Australian larval host associates include *Prospheres*, *Araucariana* (*Araucaria* – Araucariaceae), *Diadoxus* (*Callitris*, *Cupressus* – Cupressaceae; *Pinus* – Pinaceae), *Melobasis* (*Pinus*) and *Theryaxia* (*Callitris* – Cupressaceae) (Bellamy *et al.* 2013), with *Diadoxus* showing larval associations that include not only gymnosperms (Cupressaceae, Pinaceae) but angiosperms as well (Casuarinaceae, Fabaceae, Proteaceae) (Appendix 2). In addition to the above gymnosperm records, adult associations include *Astraeus* and *Diphucrania* (*Callitris*) and *Maoraxia* (*Podocarpus* – Podocarpaceae) (Appendix 2).

Collectively, these records constitute beetle taxa of diverse phylogenies.

Of the host conifer families, members of the Cupressaceae are the most widely distributed in Australia, ranging from rainforests to eucalypt-dominated open forest and semi-arid woodland. Stands of *Callitris* can be extensive in open forest and woodland, but in rainforest and wet sclerophyll forest plants are fewer and more widely separated. Many of the living Cupressaceae genera are considered Gondwanan in origin (Hill 1995), with a number of taxa recorded from Jurassic and Cretaceous deposits (Taylor *et al.* 2009). Podocarpaceae are common in Australian Tertiary deposits but with the exception of Tasmania the family is not conspicuous in the modern vegetation of Australia (Greenwood 1987), although *Podocarpus elatus* can be common in lowland littoral rainforests of the south-eastern mainland. Araucariaceae were more widely distributed and diverse in the Mesozoic but are now largely restricted to the Southern Hemisphere. Three genera (*Araucaria, Agathis, Wollemia*) occur in Australia but records of associated Buprestidae are known only from *Araucaria* (Mecke *et al.* 2005), *Araucaria* being a Gondwanan genus whose origin dates to the Late Triassic (~220 mya) and now occurring in eastern Australia, Norfolk Island, New Caledonia, New Guinea and South America. Given the rarity of the surviving populations of the single extant *Wollemia* species (*Wollemia nobilis*), and the threats posed by fire and climate change, investigation of possible buprestid–plant associations would be of particular interest.

Most buprestid host records are from angiosperms, the 'flowering plants', but unlike certain Curculionidae and Nitidulidae and others, no larval buprestids are known to develop in flowers or fruits. But like much of the fauna, flowering plants did not escape the mass extinction event that occurred at the end of the Cretaceous Period (c. 65.5 mya) and which resulted in the loss of approximately 80 per cent of known living species. Thus the Cretaceous–Tertiary extinction boundary separates a now generally extinct Cretaceous-evolved flora from one largely derived from Tertiary ancestors. How such a cataclysm, and later evolution and radiation of what constitutes a 'modern' flora, impacted on pre-existing Cretaceous plant–buprestid relationships, and the nature of the buprestid fauna as it then was, can only be guessed at. The fossil record divulges little of the fauna, and less of plant–beetle ecologies. Nevertheless, extant records hint at past ancient relationships, for the Late Cretaceous flora of the Southern Hemisphere is thought to have been less affected by the extinction event than that in the devastated Northern Hemisphere, and it is conjectured that key surviving southern groups were able to evolve and diversify (e.g. within Myrtaceae, Proteaceae [Christophel and Lys 1986; Ramírez-Barahona *et al.* 2020]), though not without secondary Tertiary extinctions, such as *Metrosideros* (Myrtaceae) from the Australian mainland (Tarran *et al.* 2016).

As with taxa associated with gymnosperms the buprestid fauna associated with angiosperms comprises species that are found on a diversity of adult and larval plant hosts – those that are larval hosts and those upon which adults are found often being quite different (Appendix 2). Host selection in wood-boring Buprestidae seems to be rather broad. Nevertheless, considering the diversity of the Australian buprestid fauna and that of the native vegetation, there are relatively few larval records (Bellamy *et al.* 2013), this largely reflecting a paucity of investigation. Among the Polycestinae, records include *Astraeus* (*Allocasuarina* – Casuarinaceae; *Bursaria* – Pittosporaceae/Bursariaceae; *Banksia* – Proteaceae; various Myrtaceae), and with *Strigoptera* providing a rare example of a species associated with mangroves (with *Camptostemon*). Regardless, surviving Australian polycestine genera demonstrate a wide range of plant families (both angiosperms and gymnosperms) of mixed lineage utilised as larval and/or adult plant hosts.

Chrysochroinae include *Austrochalcophora* (*Acacia* – Fabaceae), *Austrophorella* (*Alphitonia* – Rhamnaceae), *Chalcophorotaenia* (*Acacia* – Fabaceae), *Cyphogastra* (*Terminalia* – Combretaceae; *Elaeocarpus* – Elaeocarpaceae; *Aleurites* – Euphorbiaceae; *Acacia* – Fabaceae; *Barringtonia* – Lecythidaceae; *Alphitonia* – Rhamnaceae), *Iridotaenia* (*Mallotus* – Euphorbiaceae), *Pseudotaenia* (*Acacia* – Fabaceae). Records for several of these (e.g. *Cyphogastra*) divulge plant hosts of quite divergent phylogenies, but with species of chrysochroine *Austrochalcophora*, *Chalcophorotaenia*, *Cyphogastra* and *Pseudotaenia* utilising *Acacia* (Fabaceae) among the broader suite of known hosts.

Records for Buprestinae include genera associated with Anacardiaceae, Atherospermataceae, Araucariaceae, Bombacaceae, Casuarinaceae, Celastraceae, Cunoniaceae, Elaeocarpaceae, Euphorbiaceae, Fabaceae, Lamiaceae, Loranthaceae, Malvaceae, Myrsinaceae, Myrtaceae, Nothofagaceae, Primulaceae, Proteaceae, Rhamnaceae, Rubiaceae, Rutaceae and Sapindaceae (Appendix 2). Although the plant families cited are numerous and represent a number of evolutionary histories, given the diversity of species in the Buprestinae (and the decades of efforts towards collecting adults) larval host records are relatively few. However, there are records that indicate enigmatic mixes of seemingly ancient plant–beetle associations coupled with more geologically recent ones, as with certain *Melobasis* and *Nascioides* with Atherospermataceae, Cunoniaceae and Nothofagaceae (and *Nascioides* also with Elaeocarpaceae), but including species associated with families that are more recently evolved. Such a situation occurs within higher taxonomic ranks, for example the apparently Gondwanan tribe Epistomentini (Buprestinae) (Levey 1978b; Lawrence and Leeman 2019) in which there are taxonomically circumscribed (apparently obligate) host relationships (*Araucariana* – *Araucaria*/Araucariaceae), and with other Australian members (*Cyrioides*, *Diadoxus*) whose plant family associations are more diverse (Appendix 2). Surprisingly, known larval host records for the speciose Stigmoderinae are also few; those that exist include associations with Loranthaceae (*Castiarina*, *Temognatha*), Lamiaceae (*Castiarina*), Casuarinaceae, Rubiaceae, Sapindaceae (*Metaxymorpha*) and Myrtaceae (*Castiarina*, *Stigmodera*, *Temognatha*) (Appendix 2).

Agrilinae include taxa that have specialised plant relationships that include restriction of both larvae and adults to a particular family and genus (as with *Aaaaba*–*Rubus* – Rosaceae), and those that are almost promiscuous in the range of plant families in which larvae develop (e.g. *Agrilus*).

Plant records for adults are far more numerous but these may be less reliable as indicators of host association for, unless they are constant over time, such observations may prove no more than isolated chance encounters with an individual beetle at rest during a longer journey to and between plants unknown. This caveat aside, known anthophilous genera include all the Stigmoderinae plus the buprestine genera *Selagis* and *Neocuris*, as well as species of *Torresita* (*T. cuprifera*), *Melobasis* (*M. propinqua*), and the agriline genus *Ethonion*. Various *Diphucrania*, *Pseudanilara*, and even the usually *Callitris*- and Casuarinaceae-associated *Astraeus* (*A. hanloni*, *A. major*), may be found feeding on nectar.

The remaining buprestid genera are associated with leaves or are only encountered on fallen tree trunks or perching upon branches; and indeed so-considered larval food plant associations may be more a reflection of choice of suitable bark sculpture, or branch and trunk topography, for depositing eggs than an association based on plant phylogeny alone; the 'loose bark habitat' in the evolutionary history of certain Buprestidae is considered by Crowson (1981) a secondary return to the under-bark habitat of ancestors who had forsaken it. Of the phyllophagous species that can be reliably associated with an adult host, many species of *Melobasis* and *Diphucrania* are to be found on *Acacia* foliage and those of *Astraeus*, *Germarica*, *Dinocephalia* and *Paracephala* are well-known for perching in bright sunlight on the foliage of Casuarinaceae. Species of the taxonomically problematic *Anilara* (an unnatural grouping as currently constituted) mostly frequent the leaves of newly fallen eucalypt branches, those of *Cyrioides* are commonly associated with *Banksia* foliage, and diminutive species of *Habroloma* are to be found on the leaves of ground-hugging *Kennedia* and *Desmodium* (Fabaceae) (though also on unassociated tree species as well, such as leaves of *Elaeocarpus*).

Apart from the gymnosperm larval and adult food plant associations discussed previously the lineage of the angiosperm hosts is a diverse one, comprising those of geologically earlier and more recent origin. The Nothofagaceae (*Nothofagus*) (Fig. 537), larval hosts to certain species of *Nascioides* and *Melobasis* (both Buprestinae), are related to Northern Hemisphere Fagaceae and Betulaceae, with Jones (1986) suggesting Nothofagaceae and Betulaceae may have evolved from the same ancestor. Both families are wind-pollinated (Williams 2021) with Buprestidae playing no known role in their reproductive ecology (see discussion of wind pollination in Willmer 2011; Williams 2021). Extant Nothofagaceae are restricted to the Southern Hemisphere, inhabiting South America, Australia, New Zealand, New Guinea and New Caledonia. Fossil genera are recorded from the Upper Cretaceous (*Nothofagoxylon*) and fossil wood of *Nothofagus* is assigned to the Eocene of Australia (Grimaldi and Engel 2005). Available fossil evidence suggests *Nothofagus* and related families may have evolved in West Gondwana. In this context the disparate records of *Nascioides*–*Nothofagus* larval host associations from mainland eastern Australia, Tasmania and New Zealand (and possibly from New Caledonia) are of interest.

Casuarinaceae constitute a close relative of Nothofagaceae. The buprestid records from Casuarinaceae comprise genera (e.g. *Astraeus*, *Dinocephalia*, *Germarica*, *Nascioides*, *Paracephala*, *Stanwatkinsius*: see Bellamy *et al.* 2013) of diverse subfamily affiliation; however, the structure and unisexual nature of the flowers limits any reproductive association with anthophilous insects. Extant Casuarinaceae (4 genera, >91 spp.) are widely distributed in Australia, occurring also in East Africa, South-east Asia, Malesia, New Guinea, New Caledonia and other islands of the Pacific. Fossil Casuarinaceae are known from the Eocene (*Casuarina*) and Oligocene (described as *Gymnostoma*) of Australia (Taylor *et al.* 2009).

Atherospermataceae, historically grouped as a subfamily within Monimiaceae (Harden 1990), are restricted to the Southern Hemisphere, though they had an apparent once wider distribution. Flowers are bisexual or unisexual, with plants being either dioecious or monoecious. The family comprises 7 living genera and about 16 species (Friis *et al.* 2011). The fossil history dates to the Late Cretaceous and suggests a probable origin in West Gondwana about 100–140 mya. Despite this history published records of buprestid associates currently comprise *Nascioides* and *Melobasis* from *Doryphora* as a larval host, but the presence of other Australian Atherospermataceae genera with suitably textured bark (e.g. *Atherosperma*, *Daphnandra*) is likely to provide a wider larval host range.

Fossil Elaeocarpaceae are well known from the Southern Hemisphere, those from Australia from the Eocene onwards (e.g. Macphail *et al.* 2017). Flowers of extant species are actinomorphic and usually bisexual (Harden 1990), with the family comprising more than 600 species in 12 genera, distributed from Madagascar, east through India, South-east Asia, Japan and southern China, New Guinea and the Pacific. However, there are few Australian buprestid records for the family, despite the presence of a number of mass-flowering trees and shrubs in east coast rainforests (e.g. *Aristotelia australasica*, *Elaeocarpus grandis*, *E. holopetalus*, *E. obovatus*, *Sloanea macbrydei*, *S. woollsii*) and wet sclerophyll forests (*Elaeocarpus reticulatus*). The few records are for *Elaeocarpus* – *Habroloma* (leaves), *and Calodema* and *Metaxymorpha* (flowers), with the only larval association being that of *Cyphogastra*. The related Cunoniaceae are also well represented in fossil floras with those from Australia dated from the Late Cretaceous (*Friis et al.* 2011). Fossil flowers from the Late Eocene–Early Oligocene, and Middle Miocene, have been assigned to the extant genus *Ceratopetalum*. *Ceratopetalum* is a common floristic component of warm temperate rainforest. Flowers are actinomorphic and usually bisexual. Host records from Cunoniaceae include adult *Castiarina* (*Ceratopetalum*, *Karrabina*);

larval *Melobasis* and *Pseudanilara* (*Ceratopetalum*); adult *Maoraxia* (*Schizomeria*); larval *Cyrioides* (*Davidsonia*); and larval *Nascioides* (*Ceratopetalum*, *Schizomeria*), the majority of these being from species associated with leaves or fallen branches in rainforest.

Fossil Lauraceae date to the Middle Cretaceous with basal lineages established on either Laurasia or Gondwana by the Late Cretaceous (Chanderbali *et al.* 2001). The Lauraceae now comprise more than 2500 living species, about 130 species of which occur in Australia; most of these are endemic (Harden 1990). Flowers of Lauraceae are bisexual or unisexual. Lauraceous trees and tree-like shrubs are well represented in Australian rainforests (e.g. of *Beilschmiedia*, *Cinnamomum*, *Cryptocarya*, *Endiandra*, *Litsea*, *Neolitsea*) but there are no larval or adult hosts so far recorded. Prolonged observations in New South Wales lowland subtropical rainforests (Williams 1995; G. Williams pers. obs.) have failed to result in buprestid association records, and House (1985), in studies that included flowering *Neolitsea dealbata* and *Litsea leefeana* in tropical Queensland, also failed to note Buprestidae. Lack of larval food plant records probably suggests an absence of rearing effort from fallen timber, but the absence of anthophilous adult buprestid observations likely reflects the small size of flowers, separation of floral rewards (if any) in male and female flowers, or floral resources that are inadequately rewarding for large foraging insects.

Consisting of >80 genera, and about 1660 extant species worldwide, Proteaceae are predominantly distributed in the Southern Hemisphere with centres of diversity in Australia and southern Africa. Flowers are usually bisexual though varying from actinomorphic to strongly zygomorphic (Harden 1991), but the form and frequent bright colours of their flowers depart from the simple, open structures and white colouration theme upon which anthophilous buprestids are normally encountered. Proteaceae are thought to have originated as trees in Australia during the Middle Cretaceous (Sauquet *et al.* 2007). Eocene fossil leaves are recorded from Australia, as well as pollen grains (*Beaupreaidites*) from the Late Cretaceous, and inflorescences (*Musgraveinanthus*) from the Eocene (Taylor *et al.* 2009). The Australian flora includes genera characteristic of open forest, woodland and shrub communities (e.g. *Banksia*, *Dryandra*,[2] *Grevillea*, *Lomatia*, *Persoonia*, *Telopea* and *Hakea*). In

---

2 Species of *Dryandra* are restricted to the south-west of Western Australia. Based on molecular and morphological studies *Dryandra* was merged by Mast and Thiele (2007) with *Banksia*. Argument in support of the retention of *Dryandra*, as distinct from *Banksia*, is given by George (2008).

addition a number of genera are found in rainforest. These include *Alloxylon, Banksia, Hicksbeachia, Lomatia, Stenocarpus, Orites* and *Macadamia*. All Proteaceae exhibit structure, and often bright colour, that indicates primary adaptation to pollination by birds and/or bees, and sometimes terrestrial mammals (see discussion of floral syndromes in Williams 2021). Nevertheless, records of utilisation by buprestids (as either adult or larval food plants) are known for several plants. These include *Buckinghamia, Cardwellia, Conospermum, Darlingia, Grevillea, Hakea* and *Macadamia* (for adult *Castiarina*), *Buckinghamia, Grevillea, Macadamia* (adult *Metaxymorpha*), *Banksia* (adult and larval *Cyrioides*), *Adenanthos, Banksia, Hakea* (adult *Melobasis*), *Conospermum* (adult *Neocuris*), *Grevillea* (adult *Temognatha*), *Banksia* and *Grevillea* (adult *Diphucrania*), *Hakea* (adult *Hypocisseis*) and *Banksia, Grevillea* and *Hakea* (adult *Stanwatkinsius*).

Asteraceae worldwide include >32 000 species in >1900 genera, competing in size only with the Orchidaceae, and are mainly annual, biennial or perennial herbs and shrubs, characterised by compound inflorescences. The family is known from the Late Cretaceous of Antarctica (Barreda *et al.* 2015); however, the extensive species diversity may be of very recent geological origin (Magallón *et al.* 1999).

The Australian Asteraceae flora comprises about 1000 native species (Harden 1992) (plus >200 alien spp.), but despite this diversity there are few records of associated buprestids, these generally being for adults on flowers: *Cassinia, 'Helichrysum', Olearia, Ozothamnus (Castiarina), Ozothamnus (Melobasis), Taraxacum (Nascioides), Taraxacum (Neocuris)* and *Cassinia (Temognatha)*. These records are represented almost solely by Buprestinae. Some species (e.g. of *Cassinia, 'Helichrysum', Leucochrysum, Myriocephalus*) are quite abundant in woodlands and low heath, or along the edges of forests (as with *Cassinia straminea, Ozothamnus cassinioides*); their massed inflorescences are discernable over considerable distances, often dominating margins or as groundcover, yet commonly are devoid of buprestids. The paucity of records might partly be explained by the lack of suitable floral rewards and the inability of small flowering herbs and smaller shrub species to compete with co-flowering plants that may project stronger fragrance plumes, or that are more conspicuous in the 'floral landscape'.

Two families, the Fabaceae and Myrtaceae (Fig. 539), are popularly known for their relationship with jewel beetles, Fabaceae especially because of the numerous *Melobasis* and *Diphucrania* to be encountered on *Acacia* foliage, and Myrtaceae because of the myriad colourful *Castiarina* and related genera that seasonally frequent their fragrant and nectar-producing blossoms. Fabaceae are grouped in three subfamilies – Caesalpinioideae, Faboideae (Papilionoidae) and Mimosoidae (Harden 1991). Flowers of Caesalpinioidaeeae are slightly zygomorphic, those of Faboideae are strongly zygomorphic, but Mimosoidae have actinomorphic flowers. Caesalpinioidaeeae include 'buzz pollinated' (sonicated) shrubs of the genus *Senna* (see Williams 2021), Faboideae include *Castanospermum, Kennedia*, and *Desmodium* as well as numerous yellow-flowered 'pea flowers' (e.g. of *Bossiaea, Daviesia, Dillwynia, Phyllota, Pultenaea, Jacksonia, Oxylobium*), and Mimosoideae include *Acacia*. Fabaceae, a group with almost 20 000 living species that includes herbs, shrubs, trees and vines, are thought to have evolved during the Late Cretaceous and were already diverse and widespread by the Late Eocene (Taylor *et al.* 2009; Friis *et al.* 2011).

Worldwide the Myrtaceae are represented by about 6000 species and more than 130 genera, and today are found principally in Australia and tropical America (Taylor *et al.* 2009; Thornhill *et al.* 2015). Numerous buprestids utilise Myrtaceae as larval hosts (Hawkeswood 2002; Bellamy *et al.* 2013), with many species recorded from *Eucalyptus* (Hawkeswood 2002). Myrtaceae exemplify the association of flowering plants with anthophilous Buprestidae in landscapes dominated by open forests and woodlands, shrub communities and heathland (e.g. Hawkeswood 1978, 1980b, 1987a) (e.g. Figs 534, 544, 557, 570), but the association with many myrtaceous species (e.g. of *Backhousia, Lophostemon, Syncarpia, Syzygium, Tristaniopsis*) that inhabit rainforest and wet sclerophyll forest is less understood and studied (Williams 1995, Williams and Adam 2019; Williams 2020a, 2021). Myrtaceae are a diverse family comprising fleshy-fruited (e.g. *Syzygium*) and dry-fruited (*Tristaniopsis*) taxa (e.g. Wilson *et al.* 2001; Ladiges *et al.* 2003; Biffin *et al.* 2010). Flowers generally produce large quantities of nectar, and are actinomorphic and bisexual – but with individual species exhibiting varying degrees of self-compatibility or being self-incompatible (Williams 1995; Adam and Williams 2001; Williams and Adam 2010). The family has an extensive fossil record and is known from the Cretaceous onwards. It is represented in the Paleogene-Neogene by the fossil pollen genus *Myrtaceidites* (Thornhill and Mcphail 2012), and flowers and fruits have been reported from the Eocene of South Australia (Taylor *et al.* 2009). How the recently introduced 'myrtle rust' fungal pathogen *Austropuccinia psidii*, a serious threat to numerous Australian Myrtaceae (Williams and Adam 2019; Williams 2020a), might impact on Myrtaceae-associated Buprestidae remains to be evaluated.

BUPRESTIDS AS POLLINATORS

Ecologically specialised beetle pollination in angiosperms is thought to have evolved by the Middle to Late Cretaceous to join pre-existing guilds of beetle pollinated gymnosperms (Friis *et al.* 2011; Lawrence and Ślipiński 2013; Gottsberger 2015); with the Australian flora suggesting mutualistic interactions between beetles and flowers has been a continuous and evolving trend in angiosperms through the Tertiary (Bernhardt 2000; Williams 2021).

Although ancestral 'buprestids' are likely to have switched during the Mesozoic from gymnosperms to evolving angiosperms, their ecological role in the reproduction of pre-Tertiary angiosperms is unknown. Extant gymnosperms are insect and/or wind pollinated (cycads) or wind pollinated (conifers) (see discussion and review in Williams 2021), with living buprestids having no known part in their pollination. It is the realm of angiosperms that is the preserve of the modern pollinator buprestid fauna.

Flower-frequenting Buprestidae are known for the significant variations in abundance and diversity that can occur each year, with some groups and species noted for mass emergences punctuated by many years of absence. Such phenomena are not universal, but rather tend to be regionally unique and seemingly related to mosaic-like variation in rainfall and less so temperature (these impacting on development of immature stages and adult emergence). Flowering episodes by individual plant species and populations may also seasonally vary, their flowering fluctuating often by a number of weeks, in likely response to the same environmental cues.

Besides life history (seasonal availability of adults), behaviour (movement and feeding patterns) and morphology, the role of buprestids as pollinators is dictated fundamentally by seasonal episodes of plant flowering, flower structure and the nature of floral resources offered. Plant breeding systems (such as whether a species is self-compatible or has unisexual flowers), flower

form (size and depth of the corolla, zygomorphic or actinomorphic in structure, poricidal anthers), and the presence, absence and volume of floral rewards offered (e.g. absence of nectar in female flowers in dioecous species), constrain the plant reproduction role of visitors – invertebrate or vertebrate. In floristically complex rainforests, in which there are many flowering plant species, the wide spacing of conspecific flowering individual plants, and frequent small plant populations surviving in remnants of all types, can act as an obstacle to successful pollen movement between receptive plants. Consequently, flower visitors that show little fidelity, more correctly foraging constancy, to particular flowering plants (and thus moving unpredictably in space and time to whatever flowers are available), may poorly serve to transfer pollen between isolated individuals. In addition, it needs to be remembered that pollination, the transfer of pollen grains between individual flowers, is distinct from fertilisation. Buprestids are able to transport pollen (Williams 1995; Williams and Adam 1998), but this does not assuredly result in pollen germination and fertilisation subsequent to successful placement of pollen grains on a stigma.

There are no known instances where Buprestidae play the solitary role of pollinator. They perform this function within a suite of what is often a great assemblage of variously adapted flower visitors, not all of whom are pollinators, but which collectively compete for nectar (and sometimes pollen). In such assemblages buprestids may be abundant and diverse (as often on Myrtaceae), or seasonally and numerically rare, as on the mangrove *Avicennia marina* – Acanthaceae (Williams 2020b).

The foraging patterns of adult anthophilous buprestids are in response to seasonal flowering events and the availability of appropriate flowering food plants and the resources these provide (see also section, 'Foraging constancy'). Overall, flowering episodes commence in late winter (e.g. Fabaceae), peak in numbers of species blossoming in mid-spring to mid-summer, and then significantly reduce in number thereafter such that few plants are available in winter. Within this broad seasonal event there are marked temporal and geographical (of latitude and altitude) divergences in what plants flower where, and how individual plant species and families may respond to seasonal variation in rainfall and temperature. In the subtropical rainforests of the south-east, flowering of trees mainly occurs from October to December with few species flowering in autumn (e.g.

Alphitonia excelsa – Rhamnaceae; *Acronychia imperforata* – Rutaceae), and winter (*Cupaniopsis anacardioides* – Sapindaceae; *Synoum glandulosum* – Meliaceae), with many species and populations not flowering reliably each year (Williams 1995). And in more recent years flowering patterns for many species have become less predictable – a possible response to climate change. Although buprestids are not known to be fundamental to the reproductive ecology of flowering plants, species that flower outside of the spring–early autumn period when adult buprestids are available must rely on other invertebrates or vertebrates (or abiotic means such as obligate or facultative wind pollination). Commonly there is altitudinal variation in the flowering of related species. In the high altitude eucalypt forests of the New South Wales tablelands, for example, flowering of shrub-like *Leptospermum* commences later than that of its lowland relatives.

However, anthophilous buprestid populations may benefit from the seasonally successional flowering of food plants as has been recorded by Williams and Williams (1983) for myrtaceous *Kunzea ambigua*, *Leptospermum polygalifolium*, *Angophora hispida* and several *Eucalyptus* species flowering during spring–summer in the Sydney region of New South Wales. Overlaps in flowering can be temporally subtle. For example, on the mid-north coast of New South Wales buprestids are attracted in December to flowering *Leptospermum* growing on escarpment-associated woodland, with *Kunzea* commencing to flower a little later during January on nearby exposed volcanic peaks. Neither plant genera co-occur. Such successional phenological events can extend the supply of floral resources. But plants may also benefit by the staggered flowering patterns, the strategy reducing competition between flowering plants for the pool of potential pollinators available at any one time. The impact on buprestid populations when plants within a staggered sequence are lost (i.e. local extinction of species populations) due to habitat fragmentation, when flowering events are not seasonally reliable (as with many rainforest species), or are seasonally out of phase with buprestid emergences (due to changing climatic events), is unknown, but is an important consideration when planning for the conservation of species.

Many buprestid species exhibit restricted distribution ranges and associations with particular vegetation communities. And although many *Castiarina* seemingly show a preference for white flowers (or white flowers

just being prevalent in the 'flowering landscape'), adults of most flower-associated buprestid taxa exhibit little discrimination in the plants upon which they forage, a distinction being that the plants visited generally have actinomorphic open and shallow nectar-producing blossoms, as typified by the myrtaceous genera *Eucalyptus*, *Kunzea*, *Leptospermum* and *Syzygium*, and are highly nectiferous. Large, dish-like flowers (as with many Myrtaceae), or massed florets that may serve the same purpose, can act as small solar reflectors creating a microhabitat sympathetic to minimum energy expenditure for endogenous heat production by beetles when feeding (Heinrich and Raven 1972; Bernhardt 2000), a facility that pollinators may also seek to optimise by basking for considerable periods of time in flowers fully exposed to sunlight (as many Stigmoderinae are seen to do).

Such open, readily accessible flowers, as exemplified by Myrtaceae and other ecologically generalised mass-flowering trees and shrubs, may be the evolutionary result of pollinator-mediated selection where the occurrence of many pollination-generalist visitors imposes conflicting selection pressures and so cancels out consistent selection on specific floral traits (such as corolla form), preventing the evolution of flowers adapted for smaller guilds of ecologically specialised pollinators (Gómez *et al.* 2014).

In certain seasons in both eastern and western Australia, large numbers of *Castiarina*, *Calotemognatha*, *Temognatha* and *Stigmodera* (Stigmoderinae), and lesser numbers of *Selagis* and *Neocuris* (Buprestinae), attend to mass-flowering xeric-adapted myrtaceous genera such as *Angophora*, *Baeckea*, *Corymbia*, *Eucalyptus*, *Kunzea* and *Leptospermum*; their blossoms are a source of copious quantities of nectar (and pollen). Species of *Calodema* and *Metaxymorpha* also forage on mass-flowering Myrtaceae (including *Backhousia*, *Syzygium*, *Tristaniopsis*), both genera confined to mesic tropical and subtropical vegetation types. Members interact and compete for floral resources with a vast suite of flower-frequenting insects (e.g. Hawkeswood 1978; Williams 1995; Kitching *et al.* 2007; Williams and Adam 2019).

The iconic Australian genus *Castiarina* is widely reported as a pollinator or visitor of flowering Myrtaceae (Brooks 1948; Williams 1977; Hawkeswood 1978; Williams and Williams 1983; Williams and Adam 2019); however, the role of various *Castiarina* species in pollination and as part of flower-visiting insect assemblages has also been reported for many other plant families (Hawkeswood 2007c; Williams 2020b).

Movement of pollen

Buprestidae that visit flowers vary greatly in size. They range from small *Neocuris* (<3 mm in size) to species of *Temognatha* greater than 4 cm in length. All are capable of transporting pollen between individual flowers (e.g. Appendix 3), and individual plants within flowering populations (Williams 2020a, 2021). Movement of pollen to adjacent flowers can facilitate geitonomous pollination and fertilisation when plants are self-compatible (even if only slightly so). Xenogamous pollination results when beetles are able to transport pollen between separate flowering plants. Such inter-plant movement of pollen is necessary to achieve cross-pollination and fertilisation when a species is self-incompatible, but can result in higher levels of pollination and subsequent fertilisation where plants otherwise exhibit low levels of self-compatibility (Williams 1995, 2021; Willmer 2011). The movements of small buprestid species, and most small anthophilous insects, tend to be within the same crown and with inter-plant movements being generally to near-neighbours, though sequential 'step-wise' flight movements by small buprestids over time may progressively effect visitation by an individual to most or all flowering plants within a stand. Large Buprestidae, however, can effect pollen transfer in any single flying event over much greater distances, flight distances of greater than 100 m not being exceptional. Such relatively long-distance pollen transfer can be important in dispersing genetic material in floristically complex forests where flowering trees may be widely separated, or where an individual flowering plant is isolated from an otherwise clustered population of flowering conspecific plants. This ability by large beetles, even if numerically infrequent within the overall assemblage of flower visitors (as is the case in subtropical and tropical rainforests), can represent a significant contribution to out-crossing in plant populations and have important consequences for the reproductive success of disjunct individual plants.

In addition to behavioural patterns, such as foraging movements and contact with floral structures, the capacity of a potential pollinator to transport pollen grains and contribute to the reproductive success of a plant is related to the morphology of its body. Compounds (pollenkitt) on the outer surface, the exine, of the pollen grain, are able to facilitate adherence of grains to the vector (Faegri *et al*. 1992), but the nature of exine sculpture alone seems unrelated to the kind of pollinator involved in pollen transport (Williams and Adam 1999). Buprestidae, and flower-visiting beetles in general, were historically considered 'smooth-bodied', unable to carry significant pollen loads, and thus not likely to be effective pollinators (Archibold 1995). However, microscopic examination of buprestids collected from flowers (Williams 1995; Williams and Adam 1998) has demonstrated that individuals and species may carry heavy pollen loads, the pollen being snared by vesture and punctuation. Pollen clusters are often found adhering to mouthparts and the head more generally. Increased surface vesture and pollen-retaining body punctuation and grooves are considered derived apomorphic characters in flower-associated Coleoptera (Gardner 1989) and may have originally evolved as food collecting surfaces (Crowson 1981).

Foraging constancy

As already discussed anthophilous buprestids exhibit few instances of close pollinator–plant relationships that reflect possible evolutionary partnerships. Certain species of *Melobasis* and *Ethonion*, for example, may be reliably found on flowering Fabaceae (Hawkeswood and Turner 1994; Bellamy *et al.* 2013), but in general the relationship of beetle to plant is largely one of 'whatever flowering plants with suitable floral rewards and accessible structures' are flowering at the time – with mass-flowering species offering large amounts of nectar being favoured. Buprestid species respond to these concentrated resources only as long as they are available; it is temporal foraging constancy only in the short term, then 'switching' to the next species in flower. In this they mirror the foraging behaviour of meliphagid honey-eaters and many bees. There are no known instances of individuals trap-lining, as is recorded for certain bees and hawkmoths (Williams 2021); search patterns appear to be non-directional with little repetition of visitation to already explored flower clusters.

But few records of how pollen loads reflect 'short-term' foraging constancy are available, that of Williams and Adam (1998) providing insight into the pollen loads carried by Buprestidae in lowland rainforest remnants in northern New South Wales. Although the suite of flowering plants is limited, it nevertheless illustrates a range of plant taxa, significant pollen loads carried, and high levels of foraging constancy while plants are available.

COLOUR SECTION 1: BEETLE SPECIMENS*

Fig. 1. *Astraeus* sp. cf. *aberrans*. ©Kevin Mitchell

Fig. 4. *Astraeus fraterculus* (dorsal). ©Kevin Mitchell

Fig. 2. *Astraeus crassus*. ©Kevin Mitchell

Fig. 5. *Astraeus fraterculus* (lateral). ©Kevin Mitchell

Fig. 3. *Astraeus flavopictus*. ©Kevin Mitchell

Fig. 6. *Astraeus irregularis*. ©Kevin Mitchell

---

* Size variation within genera is discussed for each genus in Chapter 2.

Fig. 7. *Astraeus jansoni.* ©Kevin Mitchell

Fig. 10. *Astraeus oberthuri.* ©Kevin Mitchell

Fig. 8. *Astraeus navarchis* (dorsal). ©Kevin Mitchell

Fig. 11. *Helferella gothmogoides.* ©Kevin Mitchell

Fig. 9. *Astraeus navarchis* (lateral). ©Kevin Mitchell

Fig. 12. *Helferella manningensis.* ©Kevin Mitchell

Fig. 13. *Paratrachys australia.* ©Kevin Mitchell

Fig. 16. *Strigoptera bimaculata.* ©CSIRO

Fig. 14. *Paratrachys* sp. cf. *queenslandica.* ©Kevin Mitchell

Fig. 17. *Xyrocellis bumanna* (m). ©Kevin Mitchell

Fig. 15. *Polycesta mastersii.* ©Kevin Mitchell

Fig. 18. *Xyrocelis bumanna* (f). ©Kevin Mitchell

Fig. 19. *Xyroscelis crocata.* ©Kevin Mitchell

Fig. 22. *Chalcophorotaenia australasiae.* ©Kevin Mitchell

Fig. 20. *Austrochalcophora subfasciata.* ©Kevin Mitchell

Fig. 23. *Chrysodema smaragdina.* ©Kevin Mitchell

Fig. 21. *Austrophorella quadrisignata.* ©Kevin Mitchell

Fig. 24. *Cyphogastra farinosa.* ©Kevin Mitchell

Fig. 25. *Iridotaenia albivittis*. ©Kevin Mitchell

Fig. 28. *Paracupta aurofoveata*. Photographer Natalie Tees ©Australian Museum

Fig. 26. *Iridotaenia bellicosa*. ©Kevin Mitchell

Fig. 29. *Paracupta lambertii*. ©Kevin Mitchell

Fig. 27. *Lampetis foveicollis* (South Africa). ©Kevin Mitchell

Fig. 30. *Pseudotaenia ajax*. ©Kevin Mitchell

Fig. 31. *Anilara sulcipennis.* ©Kevin Mitchell

Fig. 34. *Anilara*-like genus (lateral). Photographer Natalie Tees ©Australian Museum

Fig. 32. *Anilara viridula.* ©Kevin Mitchell

Fig. 35. *Anthaxoschema terrareginae.* ©CSIRO

Fig. 33. *Anilara*-like genus (dorsal). Photographer Natalie Tees ©Australian Museum

Fig. 36. *Anthaxoschema* sp. ©Kevin Mitchell

Fig. 37. *Araucariana queenslandica.* ©Kevin Mitchell

Fig. 40. *Belionota* sp. (Philippines). ©Kevin Mitchell

Fig. 38. *Australorhipis aphanochila.* ©CSIRO

Fig. 41. *Buprestis aurulenta* (USA). ©Kevin Mitchell

Fig. 39. *Barakula petersonorum.* ©CSIRO

Fig. 42. *Buprestis novemmaculata* (Russia). ©Kevin Mitchell

Fig. 43. *Burnsiellus marmorata.* ©CSIRO

Fig. 46. *Cyrioides australis.* ©Kevin Mitchell

Fig. 44. *Chrysobothris simplicifrons.* ©Kevin Mitchell

Fig. 47. *Cyrioides imperialis.* ©Kevin Mitchell

Fig. 45. *Chrysobothris subsimilis.* ©Kevin Mitchell

Fig. 48. *Cyrioides vittigera.* ©Kevin Mitchell

Fig. 49. *Diadoxus erythrurus.* ©Kevin Mitchell

Fig. 50. *Euryspilus chalcodes.* Photographer Natalie Tees ©Australian Museum

Fig. 51. *Euryspilus viridis.* Photographer Natalie Tees ©Australian Museum

Fig. 52. *Julodimorpha saundersii.* ©Kevin Mitchell

Fig. 53. *Maoraxia auroimpressa* (m). ©Kevin Mitchell

Fig. 54. *Maoraxia auroimpressa* (f). ©Kevin Mitchell

Fig. 55. *Maoraxia* sp. cf. *storeyi*. ©Kevin Mitchell

Fig. 58. *Melobasis conica*. ©Kevin Mitchell

Fig. 56. *Melobasis azureipennis*. ©Kevin Mitchell

Fig. 59. *Melobasis cupreovittata cupreovittata*. ©Kevin Mitchell

Fig. 57. *Melobasis buprestoides*. ©Kevin Mitchell

Fig. 60. *Melobasis cupreovittata queenslandica*. ©Kevin Mitchell

Fig. 61. *Melobasis cupreovittata westralica.* ©Kevin Mitchell

Fig. 64. *Melobasis elderi.* ©Kevin Mitchell

Fig. 62. *Melobasis cupriceps.* ©Kevin Mitchell

Fig. 65. *Melobasis flexa.* ©Kevin Mitchell

Fig. 63. *Melobasis cupricollis.* ©Kevin Mitchell

Fig. 66. *Melobasis gloriosa gloriosa.* ©Kevin Mitchell

Fig. 67. *Melobasis hypocrita.* ©Kevin Mitchell

Fig. 70. *Melobasis meyricki.* ©Kevin Mitchell

Fig. 68. *Melobasis ignipicta.* ©Kevin Mitchell

Fig. 71. *Melobasis naias.* ©Kevin Mitchell

Fig. 69. *Melobasis lathami.* ©Kevin Mitchell

Fig. 72. *Melobasis nervosa.* ©Kevin Mitchell

Fig. 73. *Melobasis placida.* ©Kevin Mitchell

Fig. 76. *Melobasis splendida.* ©Kevin Mitchell

Fig. 74. *Melobasis pretiosa.* ©Kevin Mitchell

Fig. 77. *Melobasis vertebralis cuneata.* ©Kevin Mitchell

Fig. 75. *Melobasis regalis carnabyorum.* ©Kevin Mitchell

Fig. 78. *Melobasis vittata.* ©Kevin Mitchell

Fig. 79. *Merimna atrata.* ©Kevin Mitchell

Fig. 82. *Nascio vetusta.* ©Kevin Mitchell

Fig. 80. *Microcastalia globithorax.* ©CSIRO

Fig. 83. *Nascio xanthura.* ©Kevin Mitchell

Fig. 81. *Nascio chydaea* (syntype). Photographer Natalie Tees ©Australian Museum

Fig. 84. *Nascioides bicolor* (dorsal). ©Kevin Mitchell

Fig. 85. *Nascioides bicolor* (lateral). ©Kevin Mitchell

Fig. 86. *Nascioides carissimus.* ©Kevin Mitchell

Fig. 87. *Nascioides elderi.* ©Kevin Mitchell

Fig. 88. *Nascioides elessarellus* (dorsal). ©Queensland Museum, Geoff Thompson, Lily Kumpe

Fig. 89. *Nascioides elessarellus* (lateral). ©Queensland Museum, Geoff Thompson, Lily Kumpe

Fig. 90. *Nascioides macalpinei.* ©Kevin Mitchell

Fig. 91. *Nascioides* sp. cf. *nulgarra*. ©Kevin Mitchell

Fig. 94. *Nascioides quadrinotatus*. ©Kevin Mitchell

Fig. 92. *Nascioides olliffi*. ©Kevin Mitchell

Fig. 95. *Nascioides storeyi*. ©Kevin Mitchell

Fig. 93. *Nascioides parryi*. Photographer Natalie Tees ©Australian Museum

Fig. 96. *Nascioides subcostatus* (dorsal). ©Kevin Mitchell

Fig. 97. *Nascioides subcostatus* (lateral). ©Kevin Mitchell

Fig. 100. *Neocuris* sp. cf. *coerulans*. ©Kevin Mitchell

Fig. 98. *Nascioides viridis*. ©Kevin Mitchell

Fig. 101. *Neocuris* sp. cf. *crassa* (dorsal). ©Kevin Mitchell

Fig. 99. *Nascioides walfordorum*. ©Kevin Mitchell

Fig. 102. *Neocuris* sp. cf. *crassa* (lateral). ©Kevin Mitchell

Fig. 103. *Neocuris cuprilatera.* ©Kevin Mitchell

Fig. 104. *Neocuris discoflava.* ©Kevin Mitchell

Fig. 105. *Neocuris guerinii.* ©Kevin Mitchell

Fig. 106. *Notobubastes orientalis.* ©Kevin Mitchell

Fig. 107. *Notographus sulcipennis.* ©Kevin Mitchell

Fig. 108. *Pseudanilara cupripes.* ©Kevin Mitchell

Fig. 109. *Selagis adamsi.* ©Kevin Mitchell

Fig. 112. *Selagis caloptera.* ©Kevin Mitchell

Fig. 110. *Selagis atrocyanea.* ©Kevin Mitchell

Fig. 113. *Selagis corusca.* ©Kevin Mitchell

Fig. 111. *Selagis aurifera.* ©Kevin Mitchell

Fig. 114. *Selagis splendens.* ©Kevin Mitchell

Fig. 115. *Theryaxia suttoni.* ©Kevin Mitchell

Fig. 118. *Torresita* sp. 2 (m). ©Kevin Mitchell

Fig. 116. *Torresita* sp. (f). ©Kevin Mitchell

Fig. 119. *Calodema regale* (dorsal). ©Kevin Mitchell

Fig. 117. *Torresita* sp. 1 (m). ©Kevin Mitchell

Fig. 120. *Calodema regale* (lateral). ©Kevin Mitchell

Fig. 121. *Calodema regale* (ventral). ©Kevin Mitchell

Fig. 124. *Castiarina argillacea*. ©Kevin Mitchell

Fig. 122. *Calotemognatha varicollis*. ©Kevin Mitchell

Fig. 125. *Castiarina ariel*. ©Kevin Mitchell

Fig. 123. *Calotemognatha yarelli*. ©Kevin Mitchell

Fig. 126. *Castiarina armata*. ©Kevin Mitchell

Fig. 127. *Castiarina australasiae.* ©Kevin Mitchell

Fig. 130. *Castiarina burchelli.* ©Kevin Mitchell

Fig. 128. *Castiarina bremei.* ©Kevin Mitchell

Fig. 131. *Castiarina commixta.* ©Kevin Mitchell

Fig. 129. *Castiarina brutella.* ©Kevin Mitchell

Fig. 132. *Castiarina cydista.* ©Kevin Mitchell

Fig. 133. *Castiarina deliculata.* ©Kevin Mitchell

Fig. 136. *Castiarina duaringae.* ©Kevin Mitchell

Fig. 134. *Castiarina deyrollei.* ©Kevin Mitchell

Fig. 137. *Castiarina flavopurpurea.* ©Kevin Mitchell

Fig. 135. *Castiarina discoflava.* ©Kevin Mitchell

Fig. 138. *Castiarina flavoviridis.* ©Kevin Mitchell

Fig. 139. *Castiarina guttifera.* ©Kevin Mitchell

Fig. 142. *Castiarina hoffmanseggi.* ©Kevin Mitchell

Fig. 140. *Castiarina haswelli.* ©Kevin Mitchell

Fig. 143. *Castiarina jackhasenpuschi.* ©Kevin Mitchell

Fig. 141. *Castiarina hemizostera.* ©Kevin Mitchell

Fig. 144. *Castiarina kanangara.* ©Kevin Mitchell

Fig. 145. *Castiarina klugii.* ©Kevin Mitchell

Fig. 148. *Castiarina liliputana.* ©Kevin Mitchell

Fig. 146. *Castiarina latipes.* ©Kevin Mitchell

Fig. 149. *Castiarina mima.* ©Kevin Mitchell

Fig. 147. *Castiarina leai.* ©Kevin Mitchell

Fig. 150. *Castiarina nanula.* ©Kevin Mitchell

Fig. 151. *Castiarina nasuta.* ©Kevin Mitchell

Fig. 154. *Castiarina pertii.* ©Kevin Mitchell

Fig. 152. *Castiarina ocelligera.* ©Kevin Mitchell

Fig. 155. *Castiarina praetermissa.* ©Kevin Mitchell

Fig. 153. *Castiarina ochreiventris.* ©Kevin Mitchell

Fig. 156. *Castiarina quadrifasciata.* ©Kevin Mitchell

Fig. 157. *Castiarina richardsi.* ©Kevin Mitchell

Fig. 160. *Castiarina sexcavata.* ©Kevin Mitchell

Fig. 158. *Castiarina rollei.* ©Kevin Mitchell

Fig. 161. *Castiarina spectabilis.* ©Kevin Mitchell

Fig. 159. *Castiarina rudis.* Photographer Natalie Tees ©Australian Museum

Fig. 162. *Castiarina euknema.* ©Kevin Mitchell

Fig. 163. *Castiarina terminalis.* ©Kevin Mitchell

Fig. 166. *Castiarina vicina.* ©Kevin Mitchell

Fig. 164. *Castiarina* sp. cf. *tyrrhena.* ©Kevin Mitchell

Fig. 167. *Castiarina violacea.* ©Kevin Mitchell

Fig. 165. *Castiarina variopicta.* ©Kevin Mitchell

Fig. 168. *Castiarina warningensis.* ©Kevin Mitchell

Fig. 169. *Castiarina watkinsi.* ©Kevin Mitchell

Fig. 172. *Conognatha errata* (Chile). ©Kevin Mitchell

Fig. 170. *Castiarina williamsi.* ©Kevin Mitchell

Fig. 173. *Conognatha pretiossima* (Brazil). ©Kevin Mitchell

Fig. 171. *Castiarina zecki.* ©Kevin Mitchell

Fig. 174. *Hiperantha theryi* (Argentina). ©Kevin Mitchell

Fig. 175. *Metaxymorpha gloriosa* (dorsal). ©Kevin Mitchell

Fig. 178. *Metaxymorpha hauseri*. ©Kevin Mitchell

Fig. 176. *Metaxymorpha gloriosa* (lateral). ©Kevin Mitchell

Fig. 179. *Metaxymorpha imitator*. ©Kevin Mitchell

Fig. 177. *Metaxymorpha grayii*. ©Kevin Mitchell

Fig. 180. *Stigmodera cancellata*. ©Kevin Mitchell

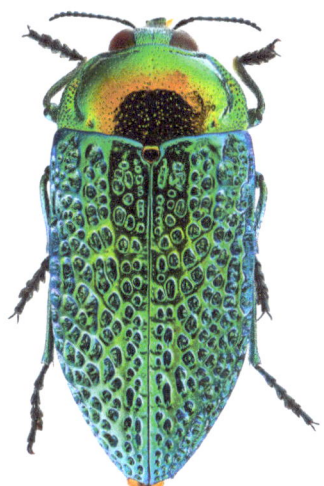

Fig. 181. *Stigmodera gratiosa.* ©Kevin Mitchell

Fig. 182. *Stigmodera macularia.* ©Kevin Mitchell

Fig. 183. *Stigmodera porosa.* ©Kevin Mitchell

Fig. 184. *Stigmodera roei.* ©Kevin Mitchell

Fig. 185. *Stigmodera sanguinosa.* ©Kevin Mitchell

Fig. 186. *Temognatha bonvouloirii* (pattern variation; see Fig. 191). ©Kevin Mitchell

Fig. 187. *Temognatha bruckii.* ©Kevin Mitchell

Fig. 188. *Temognatha carpentariae.* ©Kevin Mitchell

Fig. 189. *Temognatha chevrolatii.* ©Kevin Mitchell

Fig. 190. *Temognatha conspicillata.* ©Kevin Mitchell

Fig. 191. *Temognatha bonvouloirii* (colour variation). ©Kevin Mitchell

Fig. 192. *Temognatha duponti.* ©Robert Richardson

Fig. 193. *Temognatha gemmelli.* ©Kevin Mitchell

Fig. 196. *Temognatha heros.* ©Kevin Mitchell

Fig. 194. *Temognatha goryi.* ©Kevin Mitchell

Fig. 197. *Temognatha nigrofasciata.* ©Kevin Mitchell

Fig. 195. *Temognatha grandis.* ©Kevin Mitchell

Fig. 198. *Temognatha latithorax.* ©Kevin Mitchell

Fig. 199. *Temognatha limbata.* ©Kevin Mitchell

Fig. 202. *Temognatha obesissima.* ©Kevin Mitchell

Fig. 200. *Temognatha mnizechii.* ©Kevin Mitchell

Fig. 203. *Temognatha obscuripennis.* ©Kevin Mitchell

Fig. 201. *Temognatha murrayi.* ©Kevin Mitchell

Fig. 204. *Temognatha oleata.* ©Kevin Mitchell

Fig. 205. *Temognatha pictipes.* ©Kevin Mitchell

Fig. 206. *Temognatha rectipennis.* ©Kevin Mitchell

Fig. 207. *Temognatha rufocyanea.* ©Kevin Mitchell

Fig. 208. *Temognatha* sp. cf. *sanguineocincta.* ©Kevin Mitchell

Fig. 209. *Temognatha secularis.* ©Kevin Mitchell

Fig. 210. *Temognatha spencii.* ©Kevin Mitchell

Fig. 211. *Temognatha stevensii.* ©Kevin Mitchell

Fig. 214. *Temognatha variabilis* (fasciated). ©Kevin Mitchell

Fig. 212. *Temognatha stevensii* (fasciated). ©Kevin Mitchell

Fig. 215. *Temognatha variabilis* (partial fasciation). ©Kevin Mitchell

Fig. 213. *Temognatha thoracica.* ©Kevin Mitchell

Fig. 216. *Temognatha* sp. cf. *viridicincta.* ©Kevin Mitchell

Fig. 217. *Agrilus bispinosus.* ©Kevin Mitchell

Fig. 218. *Agrilus danesi.* ©Kevin Mitchell

Fig. 219. *Agrilus frenchi.* ©Kevin Mitchell

Fig. 220. *Agrilus hypoleucus.* ©Kevin Mitchell

Fig. 221. *Agrilus nitidus.* ©Kevin Mitchell

Fig. 222. *Agrilus walesicus.* ©Kevin Mitchell

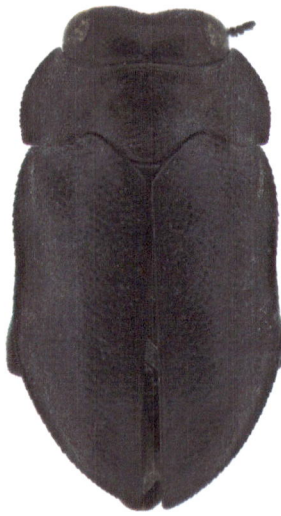

Fig. 223. *Anthaxomorphus queenslandicus* (holotype) (dorsal). Photographer Natalie Tees ©Australian Museum

Fig. 226. *Dinocephalia cyaneipennis*. ©Kevin Mitchell

Fig. 224. *Anthaxomorphus queenslandicus* (holotype) (lateral). Photographer Natalie Tees ©Australian Museum

Fig. 227. *Dinocephalia thoracica*. ©Kevin Mitchell

Fig. 225. *Aphanisticus endeloides*. ©CSIRO

Fig. 228. *Dinocephalia transsecta*. ©Kevin Mitchell

Fig. 229. *Diphucrania albosparso.* ©Kevin Mitchell

Fig. 232. *Diphucrania heroni.* ©Kevin Mitchell

Fig. 230. *Diphucrania cupripennis.* ©Kevin Mitchell

Fig. 233. *Diphucrania aurocyanea.* ©Kevin Mitchell

Fig. 231. *Diphucrania gibbera.* ©Kevin Mitchell

Fig. 234. *Diphucrania regalis.* ©Kevin Mitchell

Fig. 235. *Diphucrania roseocuprea.* ©Kevin Mitchell

Fig. 238. *Ethonion corpulentum.* ©Kevin Mitchell

Fig. 236. *Diphucrania stigmata.* ©Kevin Mitchell

Fig. 239. *Ethonion fissiceps.* ©Kevin Mitchell

Fig. 237. *Endelus* sp. (New Britain). ©CSIRO

Fig. 240. *Ethonion leai.* ©Kevin Mitchell

Fig. 241. *Ethonion* sp. cf. *maculatum*. ©Kevin Mitchell

Fig. 244. *Habroloma* sp. ©Kevin Mitchell.

Fig. 242. *Ethonion* sp. cf. *reichei*. ©Kevin Mitchell

Fig. 245. *Hedwidgiella jurecki* (South America). ©CSIRO

Fig. 243. *Germarica* sp. ©Kevin Mitchell

Fig. 246. *Hypocisseis latipennis*. ©Kevin Mitchell

Fig. 247. *Hypocisseis pilosicollis.* ©Kevin Mitchell

Fig. 250. *'Neospades'* sp. (see text, p. 94). ©Kevin Mitchell

Fig. 248. *Meliboeithon intermedium.* ©Kevin Mitchell

Fig. 251. *Pachycisseis bicolor.* ©Kevin Mitchell

Fig. 249. *Neospades chrysopygia.* ©Kevin Mitchell

Fig. 252. *Pachyschelus collaris* (Mexico). ©Kevin Mitchell

Fig. 253. *Paracephala* sp. cf. *murina*. ©Kevin Mitchell

Fig. 256. *Stanwatkinsius* sp. ©Kevin Mitchell

Fig. 254. *Sambus* sp. (dorsal) (Kenya). ©Kevin Mitchell

Fig. 257. *Synechocera deplana*. ©Kevin Mitchell

Fig. 255. *Sambus* sp. (lateral) (Kenya). ©Kevin Mitchell

Fig. 258. *Toxoscelus queenslandicus*. ©CSIRO

Fig. 259. *Trachys auriflua* (Russia). ©Kevin Mitchell

Fig. 262. *Metriorhynchus* sp. (Lycidae). ©Kevin Mitchell

Fig. 260. *Dystaxia elegans* (USA) (Schizopodidae). ©Kevin Mitchell

Fig. 263. *Palaestria* sp. (Meloidae). ©Kevin Mitchell

Fig. 261. *Chauliognathus lugubris* (Cantharidae). ©Kevin Mitchell

Fig. 264. *Morpholycus* sp. (Pyrochroidae). ©Kevin Mitchell

2 Composition, ecology and distribution of Australian genera

Subfamilies are listed systematically (following Bellamy 2002), with the re-establishment of Stigmoderinae following Buprestinae. But genera within subfamilies are listed alphabetically. Systematic placements are given in Bellamy (2002, 2003) and Lawrence and Lemann (2019). Primary references for each genus are given at the end of each listing. Host plant records are largely retrieved from Bellamy *et al.* (2013), though we have added several new records recorded by either GW or AMS since that paper. Examples of locality records for species occurring in the Gondwana Rainforests (CERRA) World Heritage Area of northern New South Wales and south-eastern Queensland are listed in Williams (2002).

Figures cited in 'bold' below refer to 'stacked focus' images; unemphasised figure references refer to *in situ* and live images.

SUBFAMILY POLYCESTINAE LACORDAIRE, 1857
Astraeus Laporte & Gory, 1837
Figs **1–10**; 265–278.
Tribe. Astraeini Cobos, 1980

General distribution. Australia but with a single known species, *Astraeus caledonicus*, occurring in New Caledonia. **Australian distribution.** Widely in Australia (~54 species) except Tasmania. Species of *Astraeus* are mainly restricted to eastern and western sections of the mainland, though a number of eastern species extend

from Queensland south to Victoria (*Astraeus dilutipes*, *A. jansoni*, *A. pygmaeus*). With the exception of *Astraeus irregularis* (occurring also in South Australia, Victoria [Bellamy 2002] and Queensland [Barker 1989]) the subgenus *Depollus* is restricted to Western Australia. Several species have apparent limited known ranges (e.g. *Astraeus aridus*, *A. mayoi*, *A. williamsi*, *A. princeps*).

Comments. Adult *Astraeus* are typically yellow and black, range in size from about 3 mm (*Astraeus pygmaeus*) to about 16 mm (*A. robustus*). Many species, especially of the subgenus *Astraeus*, are similar in pattern and size and consquently can be difficult to identify without the aid of a microscope. Adults are mainly active through late spring to mid-summer, though there is often seasonal absence of species at individual sites. Several species may at times be observed on the same tree; however, adults often employ sudden 'flicking' or 'drop' strategies to escape when disturbed.

Astraeus is the only member of the tribe Astraeini and appears to have undergone an early association with Cupressaceae and Casuarinaceae, with a later host radiation to include Proteaceae. Records commonly associate adults with species of Casuarinaceae (e.g. *Allocasuarina huegelliana*, *A. littoralis*, *A. torulosa*, *A. verticillata*, *Casuarina glauca*), in woodland, open sclerophyll forest and riparian zones, with *Allocasuarina* often as a subcanopy associate or as dense stands along sunlit forest margins, *Casuarina* forming dominant formations along coastal river banks (*C. cunninghamiana*) and semi-marshland and adjacent forest (*C. glauca*), or occasionally as discrete stands on headlands or hind dune zones (*C. equisetifolia*). Numerous species are to be found on *Callitris* foliage, with records frequently from *C. glaucophylla* (e.g. *Astraeus sundholmi*) and *C. preissii*[1] (*Astraeus crockerae*, *A. jansoni*). *Astraeus jansoni* is an example of a species frequenting both Cupressaceae and Casuarinaceae. Other plant foliage records are less common, but include those for species of *Jacksonia* (*Astraeus flavopictus*), *Hakea* and *Grevillea* (*Astraeus goldingi*), *Dryandra cirsioides* (*Astraeus meyricki*), *Acacia sclerosperma* (*Astraeus acaciae*), *Banksia prionotes* (*A. prothoracicus*), and *Banksia attenuata*, *Hakea trifurcata* and *Daviesia divaricata* (*A. fraterculus*). Records from flowering plants are infrequent but include *Astraeus hanloni* on flowering *Baeckea* sp., *Eucalyptus gracilis*, *Kunzea ambigua* and *Micromyrtus sessilis* (Myrtaceae) and *A. major* on flowering *Eucalyptus* species. (Bellamy *et al.* 2013). *Astraeus* is generally not encountered in rainforest, but a number of *Callitris* (e.g. *C. macleayana*, *C. baileyi*, *C. rhomboidea*, *C. glaucophylla*) and Casuarinaceae (*Casuarina cunninghamiana*, *C. cristata*) do occur in rainforest formations, several of which (*Callitris rhomboidea*, *C. glaucophylla* and *Casuarina cunninghamiana*) are known adult hosts where plants grow in woodland or open forest. Barker (2006c) notes the apparent rarity of *Astraeus navarchis*, owing partly to clearing of habitat, and its suspected association with *Hakea*. However, a recent specimen was collected in wet sclerophyll forest near Lansdowne in northern New South Wales, in the absence of that proteaceous genus (G. Williams unpubl. record).

Bursaria spinosa (Pittosporaceae/Bursariaceae) has been recorded as a larval host plant of *Astraeus crassus*, larvae having constructed extensive chambers in the centre of the plant wood near ground level (Hawkeswood 1984), and Hawkeswood and Peterson (1982) record *Astraeus prothoracicus* ovipositing on cones of *Banksia prionotes* (Proteaceae). In general though, larval host plant associations are otherwise poorly known.

Astraeus princeps was described by Barker (2006c) from two specimens collected from Prince of Wales

Island, Torres Strait. These were probably collected by the Australian wife (unnamed) of the insect supply dealer Robert Wind (of Monterey, California) while she was stationed there prior to World War II. The specimens, now in the South Australian Museum, Adelaide, were given to Stanley Watkins prior to his migration to Australia.

General references. Barker (1975, 1977, 1989, 1995, 1999, 2006c); Bellamy (2002, 2003); Lawrence and Lemann (2019); Turner and Hawkeswood (1996b); Williams (2002).

Helferella Cobos, 1957
Figs **11**, **12**.
Tribe. Haplostethini LeConte, 1861.

General distribution. *Helferella* is recorded from the Philippines, Indonesia, Papua New Guinea, Australia and Fiji. **Australian distribution.** Of the 10 known Australian species nine are restricted to the east coast, with individual species occurring from Tasmania to North Queensland (e.g. *Helferella frenchi*, *H. manningensis*, *H. miyal*, *H. gothmogoides*, *H. macalpinei*), and with a single species recorded from Western Australia (*H. abbreviata*). *Helferella frenchi* represents the most southern-known species (Tasmania, Victoria).

Comments. The tribe Haplostethini is widely distributed but mostly concentrated in the Americas and Africa. Non-Australian genera include *Ankareus*, *Exaesthetus*, *Mastogenius*, *Micrasta*, *Namibogenius* and *Trigonogya*. *Mastogenius* has an almost cosmopolitan distribution. Species of *Helferella* were previously placed in the subfamily Mastogeniinae, the genus characterised by minute (<3 mm), elongate and parallel-sided beetles that are normally shiny and black. Two of the Australian species (*H. abbreviata*, *H. elata*) were previously placed with *Germarica*, an agriline genus that in size and colour superficially resembles *Helferella*. *Helferella frenchi* was originally described as a species of *Mastogenius* (Théry 1928).

Adult *Helferella* are day-active, and have only been collected from foliage or occasionally in intercept traps. *Helferella gothmogoides* and *H. tolgae*, from North Queensland, have been collected at light traps. However, adults may be herbivorous. *Helferella manningensis* and *H. miyal* have been collected by beating tree foliage in littoral rainforest remnants, northern New South Wales. *Helferella manningensis* has been collected on the foliage of *Guioa semiglauca* (Sapindaceae). A single, possibly undescribed, species is known from Central Queensland (Humbolt National Park) in *Eucalyptus–Acacia* woodland (G. Williams previously unpubl. record). Larval host

---

1 *Callitris preissii*, as currently considered, is now restricted principally to the Swan Coastal Plain and Rottnest Island areas of Western Australia, and with scattered naturalised populations including the Wheatbelt, Great Southern, and Goldfields–Esperance regions. Consequently, eastern *Astraeus* plant association records for *Callitris preissii* may be attributed to *C. gracilis*.

plants are unknown; however, the related genus *Mastogenius* has been recorded in Ecuador breeding in *Acacia* and *Prosopis* (Fabaceae) and in northern America from *Quercus* (Fagaceae) and *Cercis* (Fabaceae). The larvae of the related *Ankareus somalicus* (from Somalia) breed in *Acacia* (Fabaceae).

General references. Barringer (2016); Bellamy (1990, 2002, 2003); Bellamy and Peterson (2000a); Cobos (1957); Grove and Yaxley (2004); Hołyński (1992); Lawrence and Lemann (2019); MacRae (2009); Théry (1928); Williams (1995); Williams and Weir (1987, 1988).

Paratrachys Saunders, 1873

Figs **13, 14.**

Tribe. Paratrachydini Cobos, 1980

General distribution. *Paratrachys* is a small genus known from India, South-east Asia, Indonesia, Papua New Guinea, Philippines, Japan and eastern Australia. **Australian distribution.** Two species are recorded from Australia: *Paratrachys queenslandica* from North Queensland (e.g. Atherton Tablelands) and *P. australia* from northern New South Wales.

Comments. Tribal spelling (Paratrachydini) follows Bellamy (2003). *Paratrachys* is the solitary Australian member of the tribe. However, Paratrachydini includes the non-Australian genus *Sponsor* (Africa, Madagascar, Orient).

Adult *Paratrachys* are mottled black, are somewhat elongate–ovate in appearance, and range from <2 to <3 mm in length. Adults of *Paratrachys australia* have been collected in subtropical littoral rainforest and floodplain rainforest remnants from December to March. Non-Australian species are known to be leaf-miners of *Ficus* (Moraceae) – *Helferella nigricans*, India (ex *Ficus bengalensis*) and *H. marylae* from Papua New Guinea (ex *Ficus ?wassa*) – but larval hosts for the Australian species are unknown. Although plant associations for adult *P. queenslandica* are unknown, adult *Paratrachys australia* have been consistently observed basking on the foliage of the native vine *Trophis scandens* (syn. *Malasia scandens*), a wind-pollinated member of the Moraceae. Adults readily tumble from foliage when disturbed. *Trophis scandens* occurs in numerous rainforest formations (including vine thickets) from southern New South Wales north to Central Queensland. This distribution suggests *Paratrachys australia* may be more widely distributed than its current records indicate. Hołyński (1992) considered that the known biological records for *Paratrachys*, worldwide, indicated an apparent co-evolution with various species of figs (*Ficus*, Moraceae), to which the association of *Paratrachys australia* with *Trophis* lends support.

General references. Bellamy (2002, 2003); Bellamy and Williams (1995); Evans *et al.* (2015); Hołyński (1992); Lawrence and Lemann (2019); Williams and Adam (1993).

Polycesta Seville, 1833

Fig. **15.**

Tribe. Polycestini Lacordaire, 1857

General distribution. *Polycesta* is cosmopolitan in distribution. **Australian distribution.** Only *Polycesta mastersii* is known to occur in Australia (e.g. Mourangee, Queensland).

Comments. Adults are about 15 mm in length, elongate and parallel-sided, slightly flattened in form, and grey–black with metallic reflections. *Polycesta mastersii* is associated with woodland, but appropriate habitat in Queensland is subject to ongoing broadscale clearing. Adults have been collected in November from Moreton Bay ash (*Eucalyptus tessellaris*) (E. Adams unpubl. records). Larval records for related North American and West Indian *Polycesta* include species of *Quercus* (Fagaceae), *Swietenia* (Meliaceae) and *Elaeodendron* (Celastraceae).

Polycesta and *Strigoptera* are the only members of the tribe Polycestini recorded from Australia. Worldwide, Polycestini includes about 24 genera occurring in Africa (e.g. *Aldabrica*, *Polycestina*), Madagascar (*Micropolycesta*, *Polycestaxia*), Oriental regions (*Polycestella*, *Pseudopolycesta*), Palaearctic (*Pseudocastallia*, *Thurntaxisia*) and the Neotropics (*Agenjosiana*, *Cobosesta*), with *Polycesta* widely distributed in the Northern Hemisphere.

General references. Bellamy (2002, 2003); Bellamy and Williams (1995); Lawrence and Lemann (2019); Ivie and Miller (1984); MacRae (2006).

Prospheres Saunders, 1868

Fig. 279.

Tribe. Prospherini Cobos, 1980

General distribution. Australia (1 sp.), Norfolk Island (1 sp.), New Guinea (1 sp.), New Caledonia (1 sp.). **Australian distribution.** The genus is represented in Australia by a single species, *Prospheres aurantiopictus* (syn. *Buprestis moesta* [Carter 1915]). This naturally occurs from southern Queensland to north-eastern New South Wales. Records from Victoria, South Australia, Northern Territory, Australian Capital Territory and

New Zealand refer to either adults emerging from milled timber or timber products in which the presence of live larvae had gone unnoticed, or switching to alternative hosts.

Comments. Larvae of *Prospheres aurantiopictus* are associated with *Araucaria cunninghamii* (Araucariaceae), and the natural distribution of the species is restricted to that of the host (i.e. from Macleay River, New South Wales, to New Guinea). Larvae are known to be parasitised by the small chalcidoid wasp *Oobius prospheris* (Encyrtidae). Adult beetles emerge from spring to late summer, mate and then lay eggs in the scaly bark of fallen *Araucaria cunninghamii* trees and branches. Larvae normally mine just below the surface, and develop in pupal chambers at the end of the larval tunnels. Adults have been recorded emerging at least 7 years after the milling and fashioning of timber. In addition, *Prospheres aurantiopictus* has also been recorded utilising introduced *Pinus patula* (Pinaceae) and *Acer* sp. (Sapindaceae) as larval hosts. These records represent an unusual evolutionary and ecological release from what was an apparent obligate beetle–host plant relationship with *Araucariaceae*, an obligate araucarian relationship which is exhibited by other members of *Prospheres*. *Araucaria cunninghamii* has been widely planted as an ornamental tree, and also for plantation timber, and so it is conceivable that beetles may be able to undertake limited extensions to their original restricted range, but given *Prospheres aurantiopictus* has not been recorded from other Australian members of the Araucariaceae (e.g. *Agathis robusta*, *Wollemia nobilis*), including *Araucaria bidwillii* (restricted in range to south-east Queensland but more widely planted), which naturally occurs within the geographical range of *A. cunninghamii*, the validity of records (and the ability to achieve sustainable populations) outside its natural range needs to be viewed critically.

Non-Australian members of the Prospherini include *Blepharum* (syn. *Euleptodema*, Australasian and Oriental regions) and *Hiperleptodema* (Patagonia). Levey (1978a) in his revision of *Prospheres* considered that *Prospheres* is an archaic one, originating in the Middle Cretaceous or earlier, and that the species occurring in New Caledonia, Norfolk Island and Australia had most likely been isolated from one another since the breakup of the Australian plate. *Prospheres aurantiopictus* is morphologically most similar to New Guinea *P. alternecostata*, with Levey suggesting this can be explained by the postulated colonisation of *Araucaria* from Australia sometime

between the Oligocene and Pliocene, with the time allowed for divergence between these two species being substantially less than that for *Prospheres norfolkensis* (Norfolk Island) and *P. chrysocomus* (New Caledonia).

General references. Bellamy (2002, 2003); Carter (1915); Ferrière (1947); Hawkeswood (2006a); Lawrence and Lemann (2019); Levey (1978a); Volkovitsh and Hawkeswood (1999); Williams (2002, 2020a, 2021).

Strigoptera Dejean, 1833
Fig. **16**.
Tribe. Polycestini Lacordaire, 1857

General distribution. There are 11 recognised *Strigoptera* species, these mainly occurring in the Oriental Region with two from the African region. **Australian distribution.** One species, *Strigoptera bimaculata*, occurs in northern coastal Australia.

Comments. *Strigoptera bimaculata* is distributed from India, South-east Asia to northern Australia. The species is associated with mangroves (Hawkeswood 1988), though Hawkeswood *et al.* (2018) record the species from a site in Thailand where saltwater mangroves are not present. The mangroves *Ceriops tagal* (Rhizophoraceae) and *Camptostemon schultzii* (Malvaceae/Bombacaceae) are respectively recorded as adult and larval hosts in Australia. *Camptostemon schultzii* is restricted to North Queensland, Northern Territory and the north of Western Australia but *Ceriops tagal* has a more extensive distribution, ranging from central coastal Western Australia, across coastal Northern Territory to south-eastern Queensland. Australia possesses more than 40 species of mangroves and, given this diversity and the disparate host and distributional range recorded for *Strigoptera bimaculata*, its presence at higher latitudes may not be surprising. Global warming may influence range extensions, especially if *Strigoptera bimaculata* is also able to utilise *Avicennia marina* (Acanthaceae) (the most southern of Australia's mangrove species) and other subtropical distributed species, or if host mangroves are able to extend their current western or eastern distributions in response to climate change.[2]

---

2   There has been extensive dieback of mangroves in the Gulf of Carpentaria attributed to the drying of the mangrove flats, this event believed to have been caused by climate change. See Scobell (n.d.).

General references. Bellamy (2002, 2003); Bellamy *et al.* (2013); Duke (2006); Hawkeswood (1986b, 1988); Hawkeswood *et al.* (2018); Lawrence and Lemann (2019).

Xyroscelis Saunders, 1868
Figs **17–19**; 280, 281.
Tribe. Xyroscelini Cobos, 1955
General distribution. The tribe Xyroscelini is endemic to Australia. **Australian distribution.** *Xyroscelis* is represented by *X. crocata* (south-western Western Australia) and *X. bumanna* (central coastal and inland New South Wales).

Comments. *Xyroscelis bumanna* and *X. crocata* were originally considered to be conspecific, being described by Gory and Laporte (1840) as *Amorphosoma crocatum*. Later Hope (1846) described the individual sexes as *Acmaeodera melanosticta* and *A. nodosa* (see Williams and Watkins 1986; Bellamy 2002). Hope's specimens were collected from the Swan River but the type of *A. crocatum* is simply labelled 'Nouvelle Holland'.

Locality records for *Xyrocelis crocata* include Boyagin Rock, South Perth, Warren River (near Pemberton), Wanneroo, Melville and Southern Cross. Those of *Xyroscelis bumanna* include Kurnell Peninsula (Sydney), Coonabarabran and Woy Woy (Williams and Watkins 1987). *Xyroscelis* occurs in open forests and woodland where suitable cycad hosts grow. Adults of *Xyroscelis crocata* generally occur from November to January, and into February. Adult *Xyroscelis bumanna* have been collected from November to March, but near Sydney adults seem most numerous during December. Species are restricted to cycads and may be encountered during sunny days basking on the upper surface of foliage; however, adults readily tumble from fronds when disturbed. *Xyroscelis bumanna* has been reared from the fronds of *Macrozamia communis* (Zamiaceae), and adult *X. crocata* are known to feed on the sap of *M. riedlii*, this behaviour apparently unique among Buprestidae (see *Macrozamia* Figs 536, 575).

Adults of both species are <10 mm in length, the females of *X. crocata* being somewhat larger than males, and when at rest on host leaves can be mistaken for bird droppings. *Xyroscelis* adults have a number of structural modifications suggestive of inquiline habits, or that protect beetles from foraging ants that co-occur on host foliage. They possess genal and hypomeral scrobes that receive the antennae when at rest as well as ventral and metacoxal modifications to receive the legs. The tibiae are also modified to receive the tarsi and the mouthparts are shielded by the margin of the prosternum when the head is retracted. The known host *Macrozamia* is a speciose endemic genus and although most species occur in the east of the mainland, several occur in the south-west and one species occurs in the Macdonnell Ranges of the Northern Territory. Given the wide distribution of *Macrozamia* in Australia, and the presence of other cycads (e.g. *Cycas*, *Bowenia*, *Lepidozamia*) in Australia's north and east, the continent-wide distribution of *Xyroscelis* suggests it may be more diverse than the current records indicate. The inland New South Wales records of *Xyroscelis bumanna* from Coonabarabran and that of *X. crocata* from Southern Cross (WA) lend support to this proposal.

Bellamy (1997) considered *Xyroscelis* an early divergent genus within Buprestidae, with adults superficially resembling the southern African genus *Nothomorpha* (Polycestinae–Acmaeoderini); however, in addition to certain morphological differences, adult *Nothomorpha* are flower visitors.

General references. Bellamy (1997, 2002, 2003); Lawrence and Lemann (2019); Mulder 1984; Volkovitsh and Bílý (2015); Williams and Watkins (1986).

SUBFAMILY CHRYSOCHROINAE LAPORTE, 1835
Adults of the subfamily Chrysochroinae are typically large, robust and metallic-coloured beetles. Subfamily given as Chalcophorinae in Bellamy (2002), Chrysochroinae in Bellamy (2003).

Austrochalcophora Bellamy, 2006
Fig. **20**.
Tribe. Chrysochroini Laporte, 1835
General distribution. Australia. **Australian distribution.** Comprises a single species, *Austrochalcophora subfasciata*, occurring in the Northern Territory, Queensland and Western Australia.

Comments. Adults are active in January. Adults have been recorded in Central Queensland from foliage of 'coolabah' (*Eucalyptus* sp.) and in North Queensland from foliage of *Acacia melanoxylon* (Fabaceae). Larvae have also been recorded from *Acacia melanoxylon* – this is a species chiefly distributed from Tasmania, south-east of South Australia, inland to Mt Kaputar National Park (New South Wales), and to about 200 km north of

Brisbane but with isolated populations in central and northern coastal Queensland.

Austrochalcophora subfasciata was previously placed in *Chrysodema* and *Chalcophora*; the later is a largely Holarctic genus, with most of its included species generally associated with conifers (e.g. *Pinus*, *Picea*), these hosts quite removed from the known adult and larval hosts of *Austrochalcophora subfasciata*. Bellamy (2006) considered that although the morphological differences between *Chalcophora* and *Austrochalcophora* are not so great, when taken together with the biological and biogeographical divergence from *Chalcophora* the new genus is justified.

Recorded localities include Roebourne (WA), Cape York, Innisfail, Edungalba (Qld).

General references. Bellamy (2002, 2006); Carter (1916); Hangay and Zborowski (2010); Lawrence and Lemann (2019).

Austrophorella Toyama, 1986
Fig. 21.

Tribe. Chrysochroini Laporte, 1835

General distribution. Australia. **Australian distribution.** Comprises a single species, *Austrophorella quadrisignata* occurring in New South Wales and Queensland.

Comments. *Austrophorella quadrisignata* had previously been placed in *Chalcotaenia* and *Pseudotaenia*, the species later being moved to *Austrophorella* (see Toyama 1986). Adults are active in December and January, occur in 'brigalow' woodland, and have been collected on the foliage of 'yellow jacket' (*Eucalyptus* sp.), *Alphitonia excelsa* and *A. petrei* (Rhamnaceae). Larvae have been recorded from *Alphitonia excelsa*. *Alphitonia excelsa* is a widespread tree common on the margins of various rainforest types, including littoral rainforest vine thickets, and is distributed from southern New South Wales north to Central Queensland. *Alphitonia petrei* occurs north from the mid-north coast of New South Wales in regrowth and rainforest margins. Along with other Chalcophorini *Austrophorella quadrisignata* is threatened by extensive clearing of habitat in Central Queensland and New South Wales.

Collection localities include vicinity of Duaringa, Mt Inkerman, approximately 120 km north of Clermont (Qld), Lightning Ridge, western Pilliga and Narrabri (NSW).

General references. Bellamy (2002, 2003, 2006); Bellamy *et al.* (2013); Bílý and Volkovitsh (2003);

Kerremans (1902–1903); Lawrence and Lemann (2019); Toyama (1986).

Chalcophorotaenia Obenberger, 1928
Figs **22**; 282–289.

Tribe. Chrysochroini Laporte, 1835

General distribution. Australia. **Australian distribution.** There are about 13–15 Australian species, these mainly occurring in northern and Western Australia. General distributions are *Chalcophorotaenia australasiae* (NT, WA); *C. beltanae* (NT, SA); *C. castanea* (SA, WA); *C. cerata* (NSW, Qld, SA); *C. cuprascens* Qld); *C. elongata* (Qld); *C. exilis* (SA); *C. frenchi* (Qld); *C. laeta* (Qld); *C. martinii* (WA); *C. pedifera* (Qld); *C. quadriimpressa* (NT, Qld, WA); *C. sphinx* (WA); and *C. violacea* (WA).

Comments. These are large robust and metallically coloured beetles commonly >3 cm in length. *Chalcophorotaenia cerata* and *C. laeta* flouresce brightly at night under ultraviolet light (A. Sundolm pers. obs.). Most are active during December and January. Adult *Chalcophorotaenia australasiae* have been collected at Ashburton (WA) in late February (K. and E. Carnaby pers. comm.). Species are recorded from 'mallee' (eucalypt-dominated) woodland. Adults of *Chalcophorotaenia australasiae* have been recorded from *Pleiogynium cerasiferum* (Anacardiaceae) and *Eucalyptus* spp. (Myrtaceae); *C. castanea* on *Pleiogynium cerasiferum*; and adults of *C. quadriimpressa* on foliage of *Gastrolobium grandiflorum* (Fabaceae) and *Eucalyptus* spp.

Various species have previously been placed in *Chalcophora*, *Chalcoplia*, *Chalcotaenia*, *Buprestis* and *Pseudotaenia* (Bellamy 2002, 2003, 2006).

General references. Bellamy (2002, 2003, 2006); Bellamy *et al.* (2013); Lang (2020); Lawrence and Lemann (2019).

Chrysodema Laporte & Gory, 1835
Fig. 23.

Tribe. Chrysochroini Laporte, 1835

General distribution. *Chrysodema* is essentially an Oriental genus. **Australian distribution.** The Australian fauna comprises *Chrysodema aurofoveata* (NT, Qld) and *C. furcata* (Qld). *Chrysodema aurofoveata* also occurs in New Guinea. *Chrysodema subfasciata* has been moved to *Austrochalcophora*.

Comments. Species of *Chrysodema* are moderately large (>2 cm), metallic-hued beetles. In Australia adults are active during January, but there is little published information on their ecology.

General references. Bellamy (2002); Frank and Sekerka (2020); Hołyński (2014c); Lawrence and Lemann (2019).

Cyphogastra Deyrolle, 1864
Figs **24**; 290.
Tribe. Chrysochroini Laporte, 1835
General distribution. Oriental Region, New Guinea, Australia, New Caledonia, Western Pacific. **Australian distribution.** Four species and several subspecies largely restricted to northern Australia: *Cyphogastra farinosa* and *C. pistor* (NT, Qld, WA); *C. insolens* (Cairns?); and *C. quadrivittata* (WA).
Comments. These are moderately large, metallic-hued beetles. Adults (e.g. *Cyphogastra farinosa*) are associated with rainforest, but will actively fly across intervening lightly timbered open areas.

Cyphogastra exhibits host relationships with a diversity of plants, reflecting diverse evolutionary histories. Larvae and adults of *Cyphogastra pistor* have been recorded from *Terminalia catappa* and *T. petiolaris* (Combretaceae), adults on *Pleiogynium cerasiferum* (Anacardiaceae) and *Planchonia careya* (Lecythidaceae), larvae in and adults on foliage of *Acacia bidwillii* (Fabaceae), and larvae from *Alphitonia petrei* (Rhamnaceae) and *Aleurites moluccana* (Euphorbiaceae). Larvae of *Cyphogastra farinosa* have been recorded from *Elaeocarpus angustifolius* (Elaeocarpaceae) and *Barringtona calyptrata* (Lecythidaceae). Hawkeswood's (2003) record of the non-Australian *Cyphogastra bruijni* from introduced rubber trees *Hevea brasiliensis* (Euphorbiaceae) in Papua New Guinea underscores the ecological 'plasticity' apparent in the genus.

Selected collection localities: *Cyphogastra farinosa* (Daintree River, NQld).
General references. Bellamy (2002, 2003); Bellamy *et al.* (2013); Hawkeswood (2003); Hołyński (2022); Lawrence and Lemann (2019).

Iridotaenia Deyrolle, 1864
Figs **25, 26**.
Tribe. Chrysochroini Laporte, 1835
General distribution. A large genus (~90 spp.) occurring in the Australasian, Afrotropical and Oriental regions. **Australian distribution.** Three species occur in eastern and south-eastern Australia: *Iridotaenia albivittis* (SA, NSW, Tas, Vic); *I. baumi* (NSW); and *I. bellicosa* (Qld).

Comments. The genus is a speciose one, mainly occurring in regions to Australia's north. A number of subgenera are recognised (Bellamy 2003; Hołyński 2014a). *Iridotaenia* is cited as *Scaptelytra* (Saunders 1871) in Bellamy *et al.* (2013). Here we follow Lawrence and Lemann's (2019) assignation.

Adults are moderately large (approx. 19–30 mm), elongate and dark (almost black) metallic-hued beetles with yellow/golden marginal stripes on the pronotum and elytra. Those occurring in Malesia include species that are unpatterned (e.g. *Iridotaenia igniceps*) or with distinctive metallic spots or subfasciae on the elytra (*I. quadrisignata*). Beetles are active in December and January, and can be found resting on tree trunks in eucalypt-dominated open forest. Adults of *Iridotaenia albivittis* are recorded from the foliage of *Eucalyptus* sp., and in elevated forest may be associated with *E. pauciflora* (K. Pullen pers. comm.). Larvae of *Iridotaenia bellicosa* are recorded from *Mallotus paniculatus* (Euphorbiaceae) and adults have been collected on the foliage of *Xanthostemon whitei* (Myrtaceae).

Selected collection localities: *Iridotaenia albivittis* (Rowe's Lagoon near Goulburn, vicinity of Berridale [K. Pullen pers. comm.], Sydney, Mt York [NSW]; *I. baumi*, Toronto [NSW]; *I. bellicosa*, Garradunga, Kuranda [Qld]).
General references. Bellamy (2002, 2003); Bellamy *et al.* (2013); Lawrence and Lemann (2019).

Lampetis Dejean, 1833
Fig. **27**.
Tribe. Dicercini Gistel, 1848
General distribution. Occurs in the Aftrotropical and Neotropical regions, and widely in the Northern Hemisphere. **Australian distribution.** Two likely erroneous records from Western Australia.
Comments. Carter (1924) described two *Lampetis* species, as *Buprestodes corruscans* and *B. variegata*, each based on unique specimens collected by J. Clark from Doverin and Kellerberrin (WA); erecting the genus *Buprestodes* to receive them. He described *Lampetis corruscans* (length 21–22 mm) as having the upper surface brilliant golden copper, intermixed with green, and *L. variegata* (length 19 mm) with the upper surface and sides of the elytra metallic green, the middle of the elytra violet-coppery. Carter suggested they may prove to be male and female of the same species. However, neither species has been collected since. Thus the specimens are either incorrectly labelled or are unsuccessful introductions (Bellamy 2002).

General references. Bellamy (2002, 2003); Carter (1924); Lawrence and Lemann (2019).

Metataenia Théry, 1923b

See Fig. **28**, *Paracupta aurofoveata*.

Tribe. Chrysochroini Laporte, 1835

General distribution. The genus occurs in the Australasian (including New Caledonia) and Oriental regions. **Australian distribution.** The single Australian species, *Metataenia meecki*, is recorded from Queensland, New Guinea and New Caledonia.

Comments. Hosts and habitat are unrecorded. The designation of *Metataenia meecki* (= *Paracupta meecki* Kerremans) as the type species for the genus *Metataenia* was discussed by Hołyński (2022). *Metataenia aurofoveata* (as cited in Bellamy 2002 from north-west Australia) was later transferred to *Paracupta* (Bellamy 2008a). Hołyński (2022) discusses a specimen possibly assignable to *Metataenia capitata* collected from 'S. Johnstone River', North Queensland.

Members of the genus *Metataenia* are elongate, parallel-sided beetles, in size ranging from about 17 to 23 mm. Species range from shiny black with green–purple, coppery or brassy bronze to golden green with contrasting golden depressions on the elytra and pronotum.

General References. Bellamy (2002, 2008a); Hołyński (2014d, 2022); Lawrence and Lemann (2019); Nylander (2010); Théry (1923b).

Paracupta Deyrolle, 1864

Figs **28, 29**; 291.

Tribe. Chrysochroini Laporte, 1835

General distribution. A genus of about 43 species occurring in south-east Asia, Indonesia, Papua New Guinea, Australia, and the western Pacific including New Caledonia. **Australian distribution.** Two species recorded from Australia. *Paracupta aurofoveata* occurs in north-western Australia, and *P. lambertii* from New South Wales.

Comments. *Paracupta aurofoveata* was transferred from *Metataenia*, and *P. lambertii* was previously placed in *Chalcotaenia* (Bellamy 2002, 2008a). Adults are moderately large, elongate in form, and metallic-hued. *Paracupta lambertii*, metallic bronze and ranging in size from about 18 to 25 mm, is recorded from littoral rainforest remnants and lowland subtropical rainforest in northern New South Wales. The species is active in mid to late summer. In littoral rainforest adults have been found associated with the foliage of *Elaeocarpus obovatus* (Elaeocarpaceae);

this is a medium- to large-sized tree occurring in rainforest north from central coastal New South Wales.

Selected collection localities: *Paracupta lambertii* (Iluka, Dorrigo, Harrington, Lansdowne [NSW]).

General references. Bellamy (2002, 2008a); Hołyński (2014b); Lawrence and Lemann (2019); Williams (2002, 2020a).

Pseudotaenia Kerremans, 1902–1903

Figs **30**; 292–295.

Tribe. Chrysochroini Laporte, 1835

General distribution. Australia, and Papua New Guinea (?). **Australian distribution.** Seven species recorded from Western Australia, New South Wales, South Australia and Queensland.

Comments. The adults are large (commonly >4.5 cm), metallic-hued, and have been mainly collected December–February. Habitats include mixed *Acacia–Callitris* woodland, mixed 'brigalow' and eucalypt woodland, and open forest. Extensive clearing of habitat in Central Queensland threatens a number of species. Adults are often active in daylight but are usually found resting on foliage and tree trunks. *Pseudotaenia waterhousei* has been observed during the afternoon frequenting the shaded side of (host?) trees (in Pullen 1987). *Pseudotaenia ajax* has been recorded attracted to light (ex Tambo, Central Queensland [A. Sundholm, J. Bugeja pers. obs.]). Larvae of *Pseudotaenia ajax* have been recorded from *Acacia harpophylla* (Fabaceae) and adults on *Eucalyptus cambageana, E. microtheca* and *E. moluccana* (Myrtaceae). Females preferably lay their eggs on 'injured or half dead trees' (Hobler 1925). During hot conditions adult *Pseudotaenia ajax* have been observed descending low down on the trunks of *Acacia harpophylla*. Larvae of *Pseudotaenia frenchi* are associated with *Acacia lasiocalyx* (Fabaceae) and adults with *Eucalyptus microneura*, larvae of *P. gigas* have been recorded from *Acacia acuminata, P. salamandra* adults are associated with *Casuarina* (Casuarinaceae), and larvae of *P. spilota* in *Acacia grasbyi* and adults on *Acacia cyperophylla*. Larval hosts for *Pseudotaenia waterhousei* include *Acacia doratoxylon, A. leiocalyx* (living stems) and *A. shirleyi*. Adult *Pseudotaenia waterhousei* have been observed actively chewing *Eucalyptus* leaves. Early host records for *Pseudotaenia* given as '*Casuarina*' may relate to species now placed in *Allocasuarina*.

Selected locality records: *Pseudotaenia gigas* (Coolgardie/ Wurarga, vicinity of Goomalling [WA]); *P. ajax, P. salamandra, P. waterhousei* (Duaringa [Qld]); *P. ajax*

(Warra [Qld]); *P. frenchi* (Newcastle Range/Croydon/White Mts National Park [Qld]); *P. spilota* (44 km SW Kumarina/~150 km NW Meekatharra/Shark Bay [WA]); *P. superba* (Ashburton River [WA]); *P. waterhousei* (Bimbi/Dubbo/Sandy Hollow/Mt Kaputar National Park/Round Hill Nature Reserve/Weddin Mountains/Pilliga State Forest/Lachlan/Warrumbungle Ranges [NSW]; Dunmore State Forest [Qld]); *P. salamandra* (approx. 100 km NNW Brewarrina [NSW]).

General references. Bellamy (2002, 2003, 2006); Bellamy *et al.* (2013); Bílý and Volkovitsh (2003); Lawrence and Lemann (2019); Pullen (1987).

SUBFAMILY BUPRESTINAE LEACH, 1815
Anilara Saunders, 1868
Figs **31–34**; 296, 297.
Tribe. Curidini/Curini Hołyński, 1988

General distribution. Australia and ?Papua New Guinea. **Australian distribution.** About 30 species are recognised. The genus occurs widely in Australia, though only *Anilara sulcicollis* and *A. viridula* are recorded from Tasmania.

Comments. These are mainly small coppery, bronze–black (rarely green tinged, as in *Anilara olivia*, *A. viridula*) beetles that range in size from <3 mm to about 7 mm (*Anilara sulcipennis*) in length, and somewhat flattened in form. Species are most common in open forest and woodland, usually uncommon in mesic eucalypt forest and rarely encountered in rainforest. Adults are frequently found associated with dying leaves on fallen branches of various species of *Eucalyptus* (e.g. Bellamy *et al.* 2013; Lang 2020), and in such situations can be best collected by beating with a collecting sheet or wide-mouthed net. However, when disturbed, individuals readily tumble from leaves. Adults have occasionally been observed active on the bark of fallen branches and tree trunks (e.g. *Anilara anthaxoides*), but rarely are they observed on flowers or the leaves of other plant families. One exception is *Anilara sulcipennis*, which is frequently observed basking in sunlight on the needle-like foliage of *Allocasuarina littoralis*. Larval records include *Anilara antiqua*, *A. nigrita* (*Eucalyptus crebra*), *A. convexa* (*Melaleuca decora*), *A. obscura* (*Flindersia xanthothoxyla*, *Eucalyptus* spp., *Leptospermum* sp.), *A. olivia* (*Flindersia xanthoxyla*) and *Anilara* sp. (*Flindersia maculosa*), the records from *Flindersia* representing a marked divergence in host phylogeny. *Anilara viridula* has been reared from the dead branch of an undetermined rainforest tree in subtropical rainforest (Wilson Nature Reserve, northern NSW [G. Williams unpubl. record]). Zhu *et al.* (2008) record *Anilara* associated with timber shipments from Papua New Guinea. But this may be an incorrect identification.

Carter (1926) undertook a revision of *Anilara*, in which he tabled 18 species, including the questionable record of *A. acutipennis* Théry from New Zealand, and by oversight omitting *A. pagana* Obenberger [Carter 1929]). Bellamy (2002) listed 28 species, the increase largely due to subsequent descriptions by Obenberger (1928). *Anilara* is badly in need of revision, and likely could be broken into several genera. Consequently, identification of species needs to be considered conservatively. Some species may be confused with *Anthaxoschema* and *Notographus*, but also with *Pseudanilara* and *Neocuris*. *Anilara* shares a resemblance to species of the almost cosmopolitan genus *Anthaxia* (absent from Australia), and historically several (e.g. *A. adelaidae*, *A. pagana*, *A. purpurascens*, *A. viridula*) were first placed in that genus.

A note on Australian Curidini
The Australian fauna of the tribe Curidini (as Curiini in Bellamy 2002, and Curini in Bellamy 2003), has received relatively little taxonomic attention in recent years. Australian members of its included subtribe Anilarina, as earlier constituted by Bellamy (2002), comprised the genera *Anilara*, *Anthaxoschema*, *Australorhipis*, *Barakula*, *Notographus*, *Pseudanilara*, *Theryaxia* and *Torresita*. But in his subsequent work Bellamy (2003), following Volkovitsh (2001), *Pseudanilara*, *Theryaxia* and *Torresita* were removed. Of the Australian Anilarina the genus *Anilara* in particular has remained in a problematic state, the revision by Carter (1926), and the subsequent challenging work of Obenberger (1928), leaving the status and recognition of numerous species unresolved and contentious. *Anilara* as presently constituted is a polyphyletic assemblage of species containing at least two morphologically distinct groups (Bellamy 2002), but owing to the inadequate understanding of the life history of the included taxa it is too early to state how biologically divergent a better taxonomically circumscribed definition of the subtribe may prove to be.

General references. Bellamy (1986, 2002, 2003); Bílý (2000); Burns and Burns (1992); Carter (1926); Cowie (2001); Hawkeswood (2007c); Hołyński (1988, 1993); Lawrence and Lemann (2019); Matthews (1985); Obenberger (1928); Williams (2002); Zhu *et al.* (2008).

Anthaxoschema Obenberger, 1923

Figs **35, 36**.

Tribe. Curidini Hołyński, 1988

General distribution. Restricted to Australia. **Australian distribution.** Represented by two species: *Anthaxoschema carteri* (WA) and *A. terraregenae* (Qld).

Comments. Selected collection localities: *Anthaxoschema carteri* (Peak Charles National Park, Swan River, WA). The larvae of *Anthaxoschema carteri* were collected under the bark of fallen trunks and braches of of *Acacia lasiocalyx* at Peak Charles National Park in Novemeber 2001 (Bílý and Volkovitsh 2005), but otherwise the biology and habitat of *Anthaxoschema* is not recorded.

Previously treated as a synonym of *Notographus* with which it differs in a number of characters (e.g. in *Anthaxoschema* the pronotum is widest at about the middle and only weakly constricted posteriorly, eyes only slightly convergent behind [in *Notographus* the pronotum is widest before the middle and strongly constricted behind, eyes distinctly convergent behind]) (see Lawrence and Lemann 2019). In addition to *Notographus*, *Anthaxoschema* is closely related to *Anilara*, and also additional genera such as *Barakula*, *Pseudanilara* and *Neocuris*. Larval morphology suggests that it is much closer to *Neocuris* than *Anilara* (Volkovitsh 2001; Bílý and Volkovitsh 2005).

Several series of individuals, possibly assignable to *Anthaxoschema*, have been reared from fallen *Argyrodendron actinophyllum* (Malvaceae/Sterculiaceae) branches collected in rainforest sites of northern New South Wales (Kiwarrak State Forest, Woko National Park, Tooloom Scrub) (G. Williams unpubl. records). These have the lateral margins of the pronotum widest at about the middle and only weakly constricted behind, with the posterior pronotal margin not distinctly emarginate (these characters shared with *Anilara*) but with eyes strongly convergent (as in *Notographus*).

General references. Bellamy (2002, 2003); Bellamy and Peterson (2000a); Lawrence and Lemann (2019); Théry (1945).

Araucariana Levey, 1978

Fig. **37**.

Tribe. Epistomentini Levey, 1978b

General distribution. Restricted to Australia. **Australian distribution.** Represented only by *Araucariana queenslandica*, which occurs in south-eastern Queensland.

Comments. *Araucariana* is placed in Epistomentini with *Cyrioides* and *Diadoxus* (Australia), *Cyrioxus* (New Caledonia) and *Epistomentis* (Chile, Argentina) (Bellamy 2003). Levey (1978b) considered the distribution of the tribe to be a relic one, which probably originated in the Southern Hemisphere.

Adult *Araucariana queenslandica* are about 14–16 mm in length, blackish bronze, with the punctured areas on the pronotum and elytra possessing golden reflections.

The species is known only from *Araucaria cunninghamii* (Araucariaceae), an emergent tree associated with vine thickets and subtropical rainforest. Larvae occur in fallen branches. Adults have been collected in September, November, December and January. Adults emerged in late January from a fallen narrow branch collected from the Ban Ban Range near Biggenden (south-east Qld) the preceding spring. Additional known localities include Gympie and Imbil State Forest. *Araucaria bidwillii*, or Bunya pine, also grows in south-eastern Queensland (between Gympie and the Bunya Mountains north-east of Dalby) but is not known to be a host.

General references. Bellamy (2002, 2003); Lawrence and Lemann (2019); Levey (1978b); Williams (2002, 2020a).

Australorhipis Bellamy, 1986

Fig. **38**.

Tribe. Curidini/Curini Hołyński, 1988

General distribution. Restricted to Australia. **Australian distribution.** Represented only by *Australorhipis aphanochila* (New South Wales, South Australia).

Comments. *Australorhipis aphanochila* is recorded from the vicinity of Fowlers Gap, north of Broken Hill in the arid zone western New South Wales (see Fig. 548). The type series was collected in October in ground traps (Bellamy 1986). The species has also been collected from flight intercept traps in South Australia (Lang 2020).

Species small, elongate and flattened in size (5–6 mm) and related to *Notographus*, but differs from that genus by 'having sexually dimorphic antennae; hidden labrum; eyes small, only slightly dorsally converging; tarsi as long as tibiae and pulvilli reduced' (Bellamy 1986). In addition to other distinguishing characters in *Notographus* the antennae are serrate in both sexes, and the eyes are large and strongly converging dorsally. Blackish, the elytra with dark brown areas from humeri to apices. Antennae of *Australorhipis* males are strongly pectinate, *A. aphanochila* being one of only a few Buprestidae to exhibit such a feature (others include New World *Hesperorhipis*, *Xenorhipis*, the Australian *Castiarina variegata*, and *C. shelleybarkeri* from New Guinea).

General references. Bellamy (1986, 2002, 2003); Lang (2020); Lawrence and Lemann (2019).

Barakula Peterson, 2000

Fig. **39**.

Tribe. Curidini/Curini Hołyński, 1988

General distribution. Restricted to Australia. **Australian distribution.** Represented only by *Barakula petersonorum*, and known only from Central Queensland (Barakula).

Comments. *Barakula petersonorum* is black, small (~<6 mm), elongate and flattened in form. The genus is based on a single female collected in late December 1983 in *Eucalyptus*-dominated open forest with a mixed *Allocasuarina/Callitris* understorey (Peterson and Bellamy 2000). Peterson and Bellamy opined that in relative terms *Barakula* was perhaps nearest to *Pseudanilara*. However, in general appearance and size *Barakula* resembles examples of *Maoraxia*. A number of the distinguishing features tabled in Peterson and Bellamy are shared with a long series of female *Maoraxia auroimpressa* examined by the senior author. These include a frequently exposed pygidium, eyes not converging dorsally to any significant degree and the pronotum widest in the posterior 1/3. Peterson and Bellamy (2000) note that tarsomeres 1–4 of *Barakula* each bear a short pulvillus and that pulvilli are absent on tarsomeres 1–2 in *Maoraxia*. Remnant pulvilli, however, are present on additional tarsal segments of the New Zealand *Maoraxia eremita* (Bellamy and Williams 1985; G. Williams pers. obs.). Significantly, the claws of *Barakula* are simple but appendiculate in *Maoraxia*. Nevertheless, in the absence of additional specimens *Barakula* remains somewhat enigmatic.

General references. Bellamy (2002, 2003); Lawrence and Lemann (2019); Peterson and Bellamy (2000).

Belionota Eschscholtz, 1829

Fig. **40**.

Tribe. Actenodini Kerremans, 1890

General distribution. Genus (>30 spp.) widespread in Afrotropical, Australasian, Oriental regions. Introduced to Florida (Schnepp *et al.* 2020). **Australian distribution.** *Belionota prasina* is the only recorded Australian species (King George Sound [WA]; Bloomfield River, Cape York, Endeavour River [NEQld]). This is a widespread species. Its range also includes India, China, Seychelles, South-east Asia, Indonesia and Papua New Guinea.

Comments. Largely associated with rainforest, but elsewhere also invasive of some plantation food crops. Recorded larval hosts include non-Australian *Mangifera indica* and *Anacardium occidentale* (Anacardiaceae), *Ceiba pentandra* (Bombacaceae) and *Delonix regia* (Fabaceae). More than 50 larvae were recorded infesting the trunk and branches of a single 'cashew' tree (*Anacardium occidentale*) in Goa, India (Ramasamy 2019). Adults have been recorded from *Casuarina* sp. (Casuarinaceae).

Species of *Belionota* are robust beetles, typically greater than 25 mm in length, the elytra gradually attenuate towards the apices, and the eyes large and strongly convergent. Although closely allied to *Merimna* (Melanophilini) and *Chrysobothris* (Chrysobothrini), *Belionota prasina* represents a solitary northern intrusion of a tribe widely occurring in the African, Oriental and Neotropical regions. Given the successful radiation of *Belionota* elsewhere, in particular South-east Asia and Indonesia (e.g. Philippines alone has ~8 species [Bellamy 1991]), and its apparent invasive ability to utilise plant hosts of diverse lineage, the poor representation and restricted range of the genus in Australia is surprising.

General references. Bellamy (1991, 2002, 2003); Lawrence and Lemann (2019).

Bubastes Laporte & Gory, 1836

Figs **298–301**.

Tribe. Bubastini Obenberger, 1920.

General distribution. Australia. **Australian distribution.** About 33 species, the genus occurring widely in mainland Australia, and particularly diverse in Western Australia (c. 23 spp.), Queensland (c. 11 spp.) and South Australia (c. 7 spp.) (Bílý and Hanlon 2020). *Bubastes* has not been recorded from Tasmania.

Comments. Placed with *Euryspilus* in Bubastini. *Bubastes* (syn. *Neraldus*) is largely restricted to open forest and woodland. Species are subcylindrical, parallel-sided and elongate in form, and range in size from 10 to 27 mm in length. The sexes of many species possess distinct differences in colour and size (females generally being larger), with individuals possessing a metallic-coloured lustre (e.g. *Bubastes bostrychoides* copper, *B. formosa* blue–reddish brown, *B. globicollis* black with copper reflections, *B. inconsistans* green–red, *B. sphaenoida* blue–green, *B. suturalis* green). Adults are usually active in warm weather and are encountered most frequently during late spring and summer; however, some species (e.g. *B. sphaenoida*) have been observed in late winter and early autumn. Regional patterns of emergence have been noted with adults in southern Australia generally emerging in spring, summer emergences prevailing in mid-northern Australia, and in tropical regions emergences occurring in autumn (Bílý and Hanlon 2020). Bílý and Hanlon (2020) consider that 'In low rainfall areas emergences are triggered by rain events with adults generally appearing from one to two months post rain event'. Adults of *Bubastes cylindrica* and *B. suturalis* have been recorded from foliage of *Melaleuca viridiflora* (Myrtaceae), *Bubastes vagans* has been recorded from *Melaleuca uncinata* (Myrtaceae), with other species recorded from the leaves *Eucalyptus* (Myrtaceae), *Acacia* (Fabaceae), *Allocasuarina* (Casuarinaceae) and *Xanthorrhoea* (Xanthorrhoeaceae) (Bílý and Hanlon 2020). Larval host relationships are poorly known.

General references. Bellamy (2002, 2003); Bellamy *et al.* (2013); Bílý and Hanlon (2020); Carter (1915); Lawrence and Lemann (2019); Lang (2020); Matthews (1985); Obenberger (1941).

Buprestina Obenberger, 1923

Fig. 106; also see Fig. 281 in Bellamy 2003).

Tribe. placement uncertain (see Bellamy 2003).

General distribution. Australia. **Australian distribution.** *Buprestina prosternalis* (Qld) is the only included species.

Comments. The biology and habitat of *Buprestina prosternalis* are unknown. *Buprestina prosternalis* was described by Obenberger (1923a), but owing to inconsistencies in his description of the genus and species (i.e. contrary remarks regarding the punctuation of the prosternum) Carter (1928a) considered that *Buprestina* was possibly cogeneric with *Notobubastes*. Following a comparison of specimens of *Buprestina* and *Notobubastes* in the collection of the National Museum, Prague, Bellamy and Peterson (2000a) were convinced 'that both genera should remain valid, and that they are not even particularly close'. Lawrence and Lemann (2019) consider that the two genera may not be distinct and in their key to the Australian Buprestidae the two genera are keyed out in the same couplet.

General references. Bellamy (1986, 2002, 2003); Bellamy and Peterson (2000a); Carter (1928a); Lawrence and Lemann (2019); Obenberger (1923a).

Buprestis Linnaeus, 1758

Figs **41, 42**.

Tribe. Buprestini Leach, 1815

General distribution. Genus widespread in Nearctic, Neotropical, Oriental and Palaerctic regions. **Australian distribution.** Two introduced Northern Hemisphere species recorded from New South Wales, Victoria (*Buprestis aurulenta** [Bellamy 2002]) and Western Australia (*B. novemmaculata* [Walker 2005]).

Comments. Robust, broadly elongate beetles (approx. length 17 mm) often with a metallic lustre. *Buprestis aurulenta* is a bright metallic green–red. *Buprestis novemmaculata* is black with yellow maculations on the elytra and face, and yellow lateral margins on the pronotum. In general appearance it may be confused with *Prospheres aurantiopictus*, a native species also recorded emerging from milled timber exported beyond its natural range. Larvae of *Buprestis aurulenta* are recorded overseas from *Thuja plicata* (Cupressaceae), *Abies concolor*, *A. grandis*, *Picea* sp., *Pinus contorta*, *P. flexilis*, *P. jeffreyi*, *P. lambertiana*, *P. muricata*, *P. ponderosa*, *P. radiata* and *Pseudotsuga menziesii* (Pinaceae). Larvae of *Buprestis novemmaculata* have been reared from *Abies alba*, *Larix decidua*, *Picea abies*, *Pinus halepensis*, *P. heldreichii*, *P. nigra*, *P. pinaster*, *P. pinea* and *P. sylvestris* (Pinaceae), with the record of *Buprestis novemmaculata* from Western Australia being cited as from 'pine timber' (Walker 2005). Although these host plants are not members of the Australian flora a number are grown as either plantation or ornamental trees, suggesting that introduced *Buprestis* may be able to widely establish.

**Buprestis aurulenta* was placed in *Cypriacis* by Bellamy (2002, 2003).

General references. Bellamy (2002, 2003); Lawrence and Lemann (2019); Matthews (1985); Obenberger (1923a).

Burnsiellus Levey & Bellamy, 2013

Fig. **43**.

Tribe. Buprestini Leach, 1815

General distribution. Australia. **Australian distribution.** Genus (four species) is recorded from Western Australia, Queensland, New South Wales and Victoria.

Comments. *Burnsiellus* is named after Gordon Burns, who with his wife Joy collected and studied Buprestidae expensively in Australia. The four species currently placed in the genus (*Burnsiellus albosparsa*, *B. lobatum*, *B. marmorata*, *B. trisulcata*) exhibit considerable differences from one another and might be assigned to different genera (Levey and Bellamy 2013). Adults are elongate–subcylindrical in form, moderately sized (15–20 mm), and active during spring and summer. The elytra of adult *Burnsiellus marmorata* are conspicuously patterned with yellow blotches, that of *B. albosparsa* possessing scattered small white blotches. Adult *Burnsiellus trisulcata* and *B. lobatum* are wholly black–bronze. Larval records comprise *Burnsiellus trisulcata* (NSW, SA, WA) from *Acacia* sp. (Fabaceae) and *B. marmorata* (SA, Vic) from *Melaleuca lanceolata* (Myrtaceae). *Burnsiellus lobatum* is poorly known. Levey and Bellamy (2013) suggested that the 'essentially cryptic colouration' of *B. trisulcata*, *B. albosparsa* and *B. lobatum* and the 'capture of one specimen of *B. trisulcata* at light might suggests that adults are crepuscular or nocturnal, however, Mike Powell collected a very active male specimen of *B. trisulcata* in shade at 11 am at 32 km N of Gascoyne Junction, W.A. on 2 Nov 1999'. H. Demarz collected a female in a red bucket trap in early January, on the Paynes Find Road (WA) (Levey and Bellamy 2013).

Additional collection localities: *Burnsiellus albosparsa* Cairns (Qld); *B. marmorata* Murray River (SA), Genelg River/Queenscliff (Vic), Hopetoun/Lake Hurlstone/Lake King (WA); *B. trisulcata* Bogan River (NSW), Iron Knob (SA), Paynes Find Road near Nynghan turnoff (WA).

General references. Carter (1932); Lawrence and Lemann (2019); Levey and Bellamy (2013).

Chrysobothris Eschscholtz, 1829

Figs **44, 45**.

Tribe. Chyrsobothrini Gory and Laporte, 1838

General distribution. *Chrysobothris* has an almost worldwide distribution. **Australian distribution.** There are about 15 species distributed throughout the Australian mainland. The genus is apparently absent from Tasmania.

Comments. Adult members of *Chrysobothris* are herbivores, and are not normally associated with flowers. Adults are medium-sized robust beetles (generally <2 cm in length) and are mostly encountered in December and January. Species are dark bronze–or brassy black with dull golden spots on the elytra. They can be found on foliage, especially that of *Acacia* and eucalypts, and are also encountered on freshly fallen tree trunks and branches, the females seeking fissured bark sites suitable for laying eggs. Adults of *Chrysobothris viridis* emerged from collected timber billets within 12 months of egg laying (G. Williams pers. obs.). Adults move swiftly on the surface of fallen wood, quickly moving from view and sheltering out of sight on the opposite side of fallen branches and tree trunks. Individuals have occasionally been observed at lights (Williams 1982). Adult *Chrysobothris* can be especially active during hot days. Carter (1927) quotes observations of adult *Chrysobothris mastersii* by T. G. Sloane near Young, New South Wales: 'I took over a dozen during a week in February in blazing hot weather. I got all the specimens in the bright sunshine from 11 am to 2 pm on a large fallen limb of *Eucalyptus albens*. The *Chrysobothris* was running on the rough bark of a fallen branch (had nothing to do with the leaves) and I believe lays its eggs in crevices of the bark.'

Species are mainly recorded from open forest, woodland and arid regions; however, a few (e.g. *Chrysobothris viridis*) inhabit rainforest and wet sclerophyll forest. Larvae of *Chrysobothris queenslandica* are recorded from *Acacia bidwillii* (Fabaceae), *C. saundersii* from *Allocasuarina verticillata* (Casuarinaceae) and *Acacia leiocalyx* (Fabaceae), *C. subsimilis* from *Acacia peuce* (Fabaceae) and *C. ?viridis* from *Backhousia myrtifolia* (Myrtaceae) and *Myrcine variabilis* (Primulaceae; as Myrsinaceae in earlier texts).

Collection localities include *Chrysobothris amplicollis* (Swan River [WA]), *C. caelata* (Mossgeil [NSW]), *C. mastersii* (Gayndah [Qld]), vic. Birdsville [Qld]), *C. octomaculata* (Duaringa, Charters Towers, Brisbane [Qld]), *C. perroni* (Kangaroo Is. [SA]), *C. petersoni* (Wurarga [WA]), *C. queenslandica* (Townsville [Qld]), *C. saundersi* (Rockhampton [Qld]), *C. subsimilis* (Holbrook, Mt Kaputar National Park [NSW]), *C. viridis* (Cape Hawke, Harrington, Lansdowne, Kiwarrak State Forest, Boonanghi State Forest, Iluka, Lismore [NSW]).

General references. Bellamy (2002, 2003); Bellamy *et al.* (2013); Burns and Burns (1992); Carter (1925); Cowie (2001); Hawkeswood (1986a, 1986b, 1988, 1995); Lawrence and Lemann (2019); Williams (1985, 2002, 2020a).

Cyrioides Carter, 1920

Figs **46–48**; 302.

Tribe. Epistomentini Levey, 1978b

General distribution. Australia. The related genus *Cyrioxus* occurs in New Caledonia. **Australian distribution.** Six species are recorded from Australia: *Cyrioides australis* (SA, Vic, NSW, Qld), *C. cincta* (Qld), *C. elateroides* (WA), *C. imperialis* (Tas, SA, Vic, ACT, NSW, Qld), *C. sexspilota* (Qld) and *C. vittigera* (WA).

Comments. For many years *Cyrioides* was known as *Cyria*, but this name was found to be preoccupied (Bellamy 2003). Members of *Cyrioides* are large robust beetles generally over 3 cm in length. Species are mostly yellow and black (e.g. *C. imperialis*), though it is uncertain as to whether this constitutes aposematic 'warning' colouration, but with *C. australis* being normally black. Adults are usually found on foliage or small lateral branches of their host plants. Adults normally emerge from mid-spring to autumn with peak activity in December–January; however, there is regional and seasonal variation in occurrence. Species mostly inhabit eucalypt-dominated open forest and woodland, though some (e.g. *Cyria australis*, *C. cincta*) are associated with rainforest. *Cyrioides* also includes habitat generalists, as with *Cyria australis* occurring in woodland–open forest but also in littoral rainforest, and *C. imperialis* found in woodland, open forest and heathland; their distribution is dictated by the presence of larval food plants.

Adults are mostly associated with species of *Banksia* (e.g. *Cyrioides vittigera* on *B. attenuata*, *C. australis* on *B. integrifolia*, *B. spinulosa*); however, the recorded flower or foliage associations of *C. imperialis* are more diverse and include *Angophora hispida*, *Leptospermum polygalifolium*, (Myrtaceae), *Banksia ericifolia*, *B. marginata*, *B. serrata*, and *B. spinulosa* (Proteaceae). Individuals have occasionally been recorded from flowering *Leptospermum* (Myrtaceae) (Williams 1977, Williams and Williams 1983) and *Isopogon* (Proteaceae) (A. Sundholm pers. obs.).

Known larval associations are mainly from *Banksia*. Larvae have sometimes been considered injurious. *Cyrioides imperialis* was once considered a threat to trees in Victoria (French 1900), and in a recent Tasmanian study Richards and Spencer (2018) observed that small saplings of *Banksia marginata* were vulnerable to the tunnelling activities of the same species, larval feeding tending to be confined to sapwood – though larvae had little or no effect on larger trees that were attacked. Cowie (2001) records adults of *Cyrioides imperialis* in hot weather flying like 'giant bees' among small *Banksia* bushes growing on dune and heath communities. But though conspicuous in flight beetles tend to be cryptic when resting on host plants. Individuals often undertake relatively long-distance inter-plant movements but may spend lengthy periods on food plants once landed. It is on food plants that mating usually occurs.

Collection localities: *Cyrioides australis* (Harrington [NSW]), *C. cincta* (Kuranda [Qld]), *C. elateroides* (Swan River [WA]), *C. imperialis* (Royal National Park/Blackheath/Mt York/East Minto/Coffs Harbour [NSW], Kingston, Launceston, Georgetown [Tas]), *C. sexspilota* (vic.

Innisfail/Johnstone River [Qld]), *C. vittigera* (vic. Bunker Bay, Wilga [WA]).

General references. Bellamy (2002, 2003); Bellamy *et al.* (2013); Carnaby (1987); Carter (1920); Cowie (2001); Lawrence and Lemann (2019); Williams (1977).

Diadoxus Saunders, 1868
Figs **49**; 303.
Tribe. Epistomentini Levey, 1978b

General distribution. Australia. **Australian distribution.** Three species: *Diadoxus erythurus* (NSW, Qld, SA, Vic, WA), *D. jungi* (Vic, WA) and *D. regius* (NSW, Qld, SA, Vic).

Comments. These are cylindrical, moderately small to medium-sized beetles ranging in size from about 9 to 14 mm. Adults are distinctively and cryptically patterned yellowish green, cream and black; their colour allows camouflage with host plant foliage. Peterson (in Peterson and Hawkeswood 1980) provided observations on procrypsis displayed by *Diadoxus regius* (as *D. scalaris*) associated with *Casuarina campestris* (now *Allocasuarina campestris*): individuals '... were particularly well hidden amongst foliage where they moved quickly from branchlet to branchlet usually remaining vertically orientated. They betrayed their presence only by producing a short buzzing sound with the wings and elytra immediately before taking flight. Flight was observed to be fast, straight and directed horizontally outwards from the *Casuarina* bushes.' Peterson noted that the colouration of the underside of live beetles appears to match the green colouration of the *Casuarina* branches, the disruptive colouration of the lateral margins of the beetle's abdomen further enhancing camouflage when at rest amongst foliage.

The similarly coloured, but slightly smaller species, *Diadoxus erythrurus* has been extensively studied by Hadlington and Gardner (1959). Its presence and abundance are related to that of its larval host plants (*Callitris* spp. [Cupressaceae]) but occurrence may also be influenced by seasonal rainfall patterns. However, adults have been recorded from a wider range of plant species: *Allocasuarina campestris* (Casuarinaceae), *Acacia aneura* (Fabaceae), *Grevillea eriostachya* (Proteaceae) and *Pinus* sp. (Pinaceae) (Bellamy *et al.* 2013). Larvae have been found to breed in the heartwood of older trees of *Callitris preissii* and have been reported as destructive of that species. *Diadoxus erythrurus* commonly attacks fire-damaged *Callitris*, the impact of fires creating conditions conducive to egg-laying, with eggs being deposited on the surface of bark. The life cycle is about 1 year but may be shorter (potentially just several months). Adults mainly emerge in December and January but may be encountered from September to April. Two generations can emerge each year. *Diadoxus erythrurus* and *D. regius* (syn. *D. scalaris*) are known to occur on the same plant (Hawkeswood and Turner 1997). Their larval populations may intermingle but exhibit preferential partitioning of the tree that is attacked. For example, *Diadoxus regius* has been observed to utilise the lower part of tree limbs and also to enter the root system, with *D. erythrurus* attacking the upper portion of stems and limbs (Hawkeswood and Turner 1997). Possible parasites of *Diadoxus erythrurus* have been reviewed by Hawkeswood and Turner (1997). These include wasps of the families Aulacidae, Braconidae, Cleonymidae and Eupelmidae, and also clerid beetles.

Collection localities: *Diadoxus erythrurus* (Wentworth Falls⫽46 km E of Coonabarabran⫽5 km E of Amosfield/ Moonbi lookout, N of Tamworth [NSW]), *D. jungi* (Yorke Peninsula [SA]), *D. regius* (Kellerberrin/Norseman/64 km E of Southern Cross [WA]).

General references. Bellamy (2002, 2003); Bellamy *et al.* (2013); Hadlington and Gardner (1959); Hawkeswood and Turner (1997); Lawrence and Lemann (2019); Matthews (1985); Peterson and Hawkeswood (1980); Williams and Williams (1983).

Euryspilus Lacordaire, 1857
Figs **50**, **51**; 304.
Tribe. Bubastini Obenberger, 1920
General distribution. Australia. **Australian distribution.** Four species are known: *Euryspilus australis* (WA, SA), *E. caudatus* (WA), *E. chalcodes* (WA) and *E. viridis* (WA).

Comments. Adults are metallic-hued, elongate and parallel-sided. Adult *Euryspilus viridis* have been recorded feeding on sedges (Cyperaceae). Known collection localities include *Euryspilus australis* (Port Lincoln [SA]), and *E. viridis* (Swan River [WA]).

General references. Bellamy (2002, 2003); Bellamy *et al.* (2013); Lawrence and Lemann (2019).

Julodimorpha Gemminger & Harold, 1869
Fig. **52**.
Tribe. Julodimorphini Kerremans, 1902–1903
General distribution. Australia. **Australian distribution.** There are two species: *Julodimorpha bakewelli* (SA, Vic, NSW) and *J. saundersii* (WA).

Comments. *Julodimorpha bakewelli* and *J. saundersii* are large species, with some adults being about 4.5 cm or more in length. They superficially resemble members of the Julodinae (e.g. *Amblysterna*, *Julodis*, *Sternocera*), a non-Australian subfamily occurring in African, Palaearctic and Oriental regions. However, comparative morphological studies undertaken by Bílý and Volkovitsh (2013) of the larva of *Julodimorpha* indicated a buprestine–chrysochroine affinity and that the superficial resemblance of *Julodimorpha* and Julodinae genera, with their similar life strategies, was due to convergence.

Habitats include 'mallee', heathlands and *Banksia* woodland complexes. In Western Australia adults of *Julodimorpha saundersii* have been collected in August, September and October (K. and E. Carnaby pers. obs.; Hawkeswood and Knowles 1985). Larvae have been found free-living in soil as well as in the roots of *Eucalyptus*. Adult *Julodimorpha* have been found perching on *Acacia* (Fabaceae), *Eucalyptus* and *Hysterobaeckea* (Myrtaceae) and *Dodonaea* (Sapindaceae). Females are flightless, and these are sought out by males that fly in search close to the ground (<1–2 m), mating generally being undertaken on the ground once females are found. However, during copulation males risk being attacked by ants, these being observed biting into the exposed genitalia (Gwynne and Rentz 1983). Hawkeswood and Peterson (1982) noted that a female '*Julodimorpha bakewelli*' oviposited 2.5 cm below ground in damp soil near the base of a *Calothamnus* sp. (Myrtaceae). They described a larval chamber found about 4 cm below ground surface in siliceous soil. The chamber was lined inside with matted red–brown frass, the lining created by the larva chewing a plant root that had penetrated the chamber.

An unusual behaviour observed in *Julodimorpha saundersii* (originally as *J. bakewelli*) is that of males attempting to mate with 'stubbie' beer bottles (Douglas 1980; Gwynne and Rentz 1983; Carnaby 1987), these apparently being interpreted by males as large females. Having alighted, beetles do not readily leave the bottles, persisting in their efforts to copulate. Colour reflection of the raised 'tubercles' near the base of these distinctive bottles appears to serve as an attractant and initiates sexual behaviour. Discarded beer bottles may interfere with the reproductive success of *Julodimorpha*, with the phenomenon of mating with bottles representing a behaviourally cued 'evolutionary trap' that may threaten the survival of populations (Schlaepfer *et al.* 2002; Hawkeswood 2005).

Collection localities include *Julodimorpha bakewelli* (Eyre Peninsula [SA]), *J. saundersii* (Eneabba/12 km E of Greenhead/-18 km NNW Gingin/Mount Peron [WA])

General references. Bellamy (2002, 2003); Bellamy and Weir (2008); Bellamy *et al.* (2013); Bílý and Volkovitsh (2013); Hawkeswood (2005); Hawkeswood and Peterson (1982); Lawrence and Lemann (2019).

Maoraxia Obenberger, 1937
Figs **53–55**.
Tribe. Maoraxiini Hołyński, 1984

General distribution. Philippines, Australia, Lord Howe Island, Fiji, Tonga and New Zealand. **Australian distribution.** Two described mainland species: *Maoraxia auroimpressa* (NNSW, CQld) and *M. storeyi* (NQld). A third described species, *Maoraxia roseocuprea*, is endemic to Lord Howe Island (Bellamy and Peterson 2000b).

Comments. The genus was first described as *Maoriella* (Obenberger 1924b) but this name was found to be preoccupied (Bellamy and Williams 1985). The presence of a prominent deeply bilobed tarsal pad (pulvillus) on the 4th tarsomere, and the general absence of pulvilli on tarsomeres 1–3, is a character unique to *Maoraxia*, though as discussed earlier under *Barakula* remnant tarsal pads are to be found on some segments of the New Zealand species *Maoraxia eremita* (Bellamy and Williams 1985). Tarsal claws are appendiculate, a feature absent in related genera such as *Neocuris*, *Pseudanilara*, *Barakula* and *Notographus*. Adults are small (3–5 mm), with females somewhat larger than males, and are active in late spring and summer. Individuals are normally found on sunlit foliage. *Maoraxia auroimpressa* occurs widely in littoral rainforest situated along the eastern seaboard of the mainland (at least as far north as Eurimbula National Park in central coastal Queensland), as well as hinterland lowland subtropical rainforest and dry rainforest. *Maoraxia storeyi* is recorded from North Queensland littoral rainforest and hinterland montane rainforest. A male specimen possibly assignable to *Maoraxia storeyi* has been collected from 'vine scrub' in the vicinity of Many Peaks and Calliope, south-east Queensland (G. Williams pers. obs.).

Recorded larval host plants for *Maoraxia auroimpressa* include *Elaeodendron australe* (Celastraceae), *Alectryon coriaceus* and *Guioa semiglauca* (Sapindaceae); with females favouring fissured branches of narrow width (<4 cm) upon which to lay their eggs. Adults are recorded from the foliage of *Elaeodendron australe* (Celastraceae),

Podocarpus elatus (Podocarpaceae), *Acronychia* sp. (Rutaceae), *Schizomeria ovata* (Cunoniaceae), *Syzygium floribundum*, *S. smithii* (Myrtaceae) and *Argyrodendron actinophyllum* (Malvaceae) but are likely to frequent a much larger suite of plants. The hosts of *Maoraxia storeyi* are unknown; the species has been collected at lights (Bellamy 1991).

Adults of *Maoraxia auroimpressa* differ significantly in size and colour, males being about 3 mm in length and largely black, and females about 4.5 mm and variably shiny blue to blue–green.

Selected collection localities: *Maoraxia auroimpressa* (Hallidays Point/Redhead/Manning Point/Harrington/vic. Lansdowne/Kiwarrak SF/Camden Head/Kerewong SF/Iluka/Wilson Nature Reserve [NSW], Wide Bay/Eurimbula National Park [SE-CQld]); *M. storeyi* (Conway Range/Kuranda/Cape Tribulation [NQld]).

General references. Bellamy (1991, 2002, 2003); Bellamy and Peterson (2000a, 2000b); Bellamy and Williams (1985); Bellamy *et al.* (2013); Bílý and Volkovitsh (2005); Hołyński (1984, 1988); Lawrence and Lemann (2019); Williams (2002, 2020a).

Melobasis Gory & Laporte, 1837
Figs **56–78**; **305–311**.
Tribe. Melobasini Bílý, 2000

General distribution. Laos, Malaysia, Indonesia, Australia, New Caledonia, Solomon Islands. **Australian distribution.** More than 200 species and subspecies recorded from Australia (including Tasmania and Torres Straits Islands), Lord Howe Island and Norfolk Island; however, very few species are widely distributed (Levey 2012).

On the basis of known distribution Levey (2012) identified five areas of high endemism and two areas of low endemism. The areas of high endemism are all peripheral, suggesting *Melobasis* is part of a wet-adapted fauna. Inland Australia and non-arid South Australia and north-western Victoria appear to be areas of low endemism, with most species occurring elsewhere and a number being widespread. The fauna is largely derived from peripheral areas of south-west and south-east Australia and probably represents a fauna better adapted to arid and semi-arid conditions. South-western Australia is rich in species (82), of which 60 are endemic. Forty-nine species are recorded from the humid and subhumid zones of south-east Australia (to southern Queensland), 31 of which are endemic and several are associated with rainforest (e.g. *Melobasis conica, M. hypocrita, M. ignipicta, M. elderi, M. williamsi*). Very few of these are shared to North Queensland. Levey (2012) considered Tasmania and the highland areas of Victoria and New South Wales to be relatively poor in species, however, Cowie (2001) lists 12 species from Tasmania. Twenty-five species are known from subhumid Central Queensland, this region coextensive with the North and South Brigalow Belt bioregions (Levey 2012). Twenty-two species are recorded from the Queensland tropical rainforest areas south of Cape York Peninsula; however, owing to a lack of detailed collection data it is not known how many are true rainforest species. Twenty species are known from the northern Tropical region, 17 of which are endemic, but this is a poorly collected area and the currently recorded fauna may be an understatement of the region's *Melobasis* diversity.

Comments. Levey (2012) considered the closest relatives to *Melobasis* to be *Merimna, Pseudanilara, Melanophila, Cylindrophora, Anthaxioides, Trachypteris* and *Juniperella*, the last five genera not occurring in Australia. The genus consists of small- to moderately sized (4–8 mm), metallic green–bronze beetles. Many possess distinctly patterned elytra, and often exhibit significant variation in pattern between subspecies (as in those of *Melobasis cupreovittata*), or between sexes (e.g. *M. gloriosa*) (Levey 2012, 2018). *Melobasis* has occasionally been observed at lights (Williams 1982); adults typically can be found perching on stems and foliage in bright sunlight. All species are day-active, with adults being mainly encountered in spring and summer. Rarely are adults seen in cooler months, though a number of Western Australian species (e.g. *Melobasis regalis carnabyorum, M. eximia*) have been recorded in autumn–mid-year. Species occupy a diversity of habitats, ranging from montane cool temperate rainforest (e.g. *Melobasis hypocrita, M. ignipicta*) to coastal dune-associated closed shrub complexes (*M. anchoralis, M. naias*). Although many species are known, at a coarse habitat scale, only from particular vegetation formations (i.e. woodland, rainforest, heathland), others are vegetation community generalists, their occurrence dictated by the presence or absence of larval food plants rather than the overarching plant community in which these grow. *Melobasis hypocrita* and *M. ignipicta*, for example, cited above as occurring in cool temperate rainforest (where *Nothofagus* is a recorded larval host), also are known from lower altitude and more xeric-adapted plant communities (e.g. see Burns and Burns 1992).

Although there are many records of adults associated with foliage (this possibly also being that of the larval host), and especially that of *Acacia* (Bellamy *et al.* 2013), there are few flower-associated records; an example is *Melobasis propinqua propinqua* eating petals (J. R. Turner 1984, cited in Levey 2012). It is likely that all species live as larvae in woody or semi-woody tissues of vascular plants (Levey 2012). Larval records include *Melobasis abigailae* (*Acacia spectabilis* [Fabaceae]), *M. abnormis* (*Santalum acuminatum* [Santalaceae]), *M. apicalis* (*Bossiaea rhombifolia, Jacksonia scoparia* [Fabaceae]), *M. conica* (*Argyrodendron actinophyllum* [Malvaceae]), *M. costata* (*Acacia harpophylla*), *M. cupriceps* (*Acacia longifolia, Jacksonia scoparia, Oxylobium aciculiferum* and *Viminaria juncea* [Fabaceae]), *M. fulgurans* (*Acacia dealbata*), *M. goerlingi* (*Acacia* sp.), *M. hypocrita* (*Doryphora sassafras* [Atherospermataceae], *Nothofagus moorei* [Nothofagaceae, in error as Fagaceae in Bellamy *et al.* 2013], *Corymbia tessellaris* [Myrtaceae], *Pinus radiata* [Pinaceae], *Ceratopetalum apetalum* [Cunoniaceae]), *M. ignipicta* (*Nothofagus moorei*), *M. jacquelinae* (*Jacksonia scoparia*), *M. kaszabi* (*Alphitonia petriei* [Rhamnaceae]), *M. naias* (*Acacia longifolia, A. mucronata*), *M. nervosa* (*Acacia dealbata, A. mucronata, A. sophorae*), *M. obscurella* (*Acacia mucronata*), *M. propinqua* (*Pultenaea villosa* [Fabaceae]), *M. purpurescens* (*Acacia concurrens, A. dealbata, A. irrorata, A. longifolia, A. melanoxylon, A. mucronata, A. sophorae, A. terminalis, Flindersia australis* [Rutaceae/Flindersiaceae], *Citrus limonia, C. sinensis, Citrus* sp.[Rutaceae], *Clerodendrum floribundum* [Verbenaceae]), *M. pusilla* (*Acacia mearnsii*), *M. sexplagiata* (*Eucalyptus camaldulensis, E. rudis* [Myrtaceae]), *M. splendida* (*Acacia longifolia, Eucalyptus macrorhyncha, E. punctata*), *M. vertibralis* (*Acacia leiocalyx*) and *M. vittata* (*Acacia chalkeri, A. cyclops*) (Bellamy *et al.* 2013; G. Williams unpubl. records). Interestingly, the mangroves *Excoecaria agallocha* (Euphorbiaceae) and *Aegiceras corniculatum* (Myrsinaceae) are recorded as larval hosts of *Melobasis purpurescens* (Hockey and De Baar 1988; Bellamy *et al.* 2013) (mangroves generally are a poorly collected habitat).

General references. Bellamy (2002, 2003); Bellamy *et al.* (2013); Bílý (2000); Burns and Burns (1992); Carter (1923b); Cowie (2001); Hockey and De Baar (1988); Lawrence and Lemann (2019); Levey (2012, 2018, 2023); Williams (1985, 2002, 2020a).

Merimna Saunders, 1868
Fig. 79.
Tribe. Melanophilini Bedel, 1921

General distribution. Australia, New Guinea. **Australian distribution.** Represented only by *Merimna atrata*, commonly referred to as the 'fire beetle'. Widespread, recorded from all mainland states. Absent from Tasmania.

Comments. *Merimna atrata* is a medium-sized nocturnal species (~17–27 mm), completely dusky black, and was originally described as a species of *Chrysobothris*. *Chrysobothris* can be distinguished by the presence of golden (if sometimes vague, as in *C. subsimilis*) impressions on the elytra whereas the elytra of *Merimna* are wholly black. Adult *Merimna* have been recorded from spring to autumn and exhibit distinctive behaviour in that they are attracted to a range of light sources (including UV, mercury vapour and domestic flourescent) and to recently burnt timber. In nature adult *Merimna atrata* approach forest fires because their larvae develop in freshly burnt wood. To facilitate this attraction adults possess two pairs of specialised infrared absorbing receptor organs on the ventrolateral sides of the abdomen (Schmitz *et al.* 2001; see also Schneider and Schmitz 2013) – warming causes an increase in receptor activity. Results of a study by Hinz *et al.* (2018) suggest that the beetle's infrared receptors serve as an early warning system preventing accidental landing on a hot surface. *Merimna atrata* has also been observed in numbers attracted to hot brick walls (Nullabor Roadhouse, Western Australia), and also campfires (A. Sundholm, J. Bugeja pers. obs.).

Known habitats include woodland and eucalypt-dominated open forest but beetles are difficult to see during the day as they tend to rest inconspicuously (or quickly move from sight) on the dark trunks and branches of sheltering trees. Females have been observed laying eggs into smouldering bark at the base of *Corymbia calophylla* (as *Eucalyptus calophylla*) (Hawkeswood 2007a), with additional larval food plants including myrtaceous *Baeckea frutescens*, *Eucalyptus albens*, *E. melliodora* and/ or *E. moluccana*. Adults have only rarely been recorded on flowers, an instance being that from flowering *Angophora hispida* (Williams and Williams 1983). There is an interesting record in Carter (1927), quoting information from T. G. Sloane, of *Merimna atrata* capturing and devouring a small *Anilara* buprestid.

Selected collection localities: Sydney (NSW), vic. Timber Creek/Katherine Gorge/Taylor Creek (NT), Barcaldine/Cudmore National Park/Jardine River/Palm Cove (Qld), Perth (WA).

General references. Bellamy (2002, 2003); Bellamy *et al.* (2013); Burns and Burns (1992); Hawkeswood (2007a); Hawkeswood and Peterson (1982); Kitchen (2009); Lawrence and Lemann (2019); Matthews (1985); Williams (1982).

Microcastalia Heller, 1891
Fig. **80.**
Tribe. Bubastini?
General distribution. Australia. **Australian distribution.** Two species: *Microcastalia globithorax* (SWWA, SA, Vic, SEQld), *M. scintillans* (SA, southern WA).

Comments. Both species were originally placed in the genus *Castalia* (Bellamy 2002). Adult *Microcastalia globithorax* range in size from ~12 mm (males) to ~14 mm (females), with adult *Microcastalia scintillans* being about 14 mm in length. Individuals are recorded from open woodland, and adults are generally active from November through to February. Larvae of *Microcastalia globithorax* have mainly been recorded from green stems of the hemiparasitic shrubs *Choretum chrysanthum* and *C. glomeratum* (Santalaceae), with an additional record from *Leptomeria aphylla* (Santalaceae) (Lang and Stolarski 2020). Adults have been recorded on the foliage of *Choretum glomeratum*. Although the larval hosts of *Microcastalia scintillans* are unknown Lang and Stolarski (2020) considered that 'host specificity, endemicity and comparable ranges of *Microcastalia* and *Choretum* are consistent with having evolved together in Australia'. More broadly, the known larval host associations (*Choretum*, *Leptomeria*) suggest a possible evolutionary specialisation with Santalaceae, a cosmopolitan family with 10 genera and about 50 species recorded from Australia (Hnatiuk 1990).

Selected locality records: *Microcastalia globithorax* (10 km N of Leeman [WA], Kangaroo Island/Eyre Peninsula [SA]), *M. scintillans* (approx. 31 km NW of Port Augusta/Eyre Peninsula [SA], Kellerberrin [WA]).

General references. Bellamy (2002, 2003); Bellamy and Peterson (2000a); Lang and Stolarski (2020); Lawrence and Lemann (2019).

Nascio Laporte & Gory, 1837
Figs **81–83.**
Tribe. Nascionini Hołyńki, 1988.
General distribution. Australia.
Australian distribution. Four species: *Nascio chydaea* (WA), *N. simillima* (NSW, Qld), *N. vetusta* (SA, Vic, NSW, Qld) and *N. xanthura* (NSW).

Comments. The genus exhibits a marked distribution disjunction with three species occurring in eastern mainland Australia, and a single species restricted to Western Australia. Adults are elongate in form, range in size from approximately 10 mm (*Nascio xanthura*) to about 17 mm (*N. vetusta*) in length, and are variably dull bronze–black with brownish yellow markings on the elytra (these restricted in *N. xanthura* to the apical third); the cryptic appearance produced allows camouflage when at rest. Most species emerge in late spring–early summer, but may be present in late summer. The adults can be very active in sunlight, occasionally seen moving rapidly on tree trunks. Habitats include dry sclerophyll forest, wet sclerophyll forest and woodland. Larvae of *Nascio simillima* are recorded from *Eucalyptus drepanophylla*, and adults have been found on foliage of *E. resinifera* and other *Eucalyptus* species. Larval host plants for *Nascio vetusta* include *Corymbia tessellaris*, *Eucalyptus baxteri*, *E. goniocalyx*, *E. haemastoma*, *E. obliqua*, *E. saligna* and *Metrosideros* sp. (Myrtaceae). Adult *Nascio xanthura* have been observed resting on the branches of *Allocasuarina* (Casuarinaceae) in open sclerophyll forest, and coastal woodland in proximity to the seaside (G. Williams pers. obs.).

Selected collection localities: *Nascio simillima* (Gowlerstown/Lansdowne/Mt Warning [NSW], Ravenshoe (Qld); *N. vetusta* (Engadine/Wyong [NSW]); *N. xanthura* (Yarratt State Forest/Hat Head [NSW]); *N. chydaea* (Salt River [WA]). See Peterson (2018) and our discussion of bupestrids in the section, 'South-west Western Australia' for a clarification of the type locality of *Nascio chydaea*.

General references. Bellamy (2002, 2003); Bellamy *et al.* (2013); Lawrence and Lemann (2019); Williams (2002); Williams and Watkins (1985).

Nascioides Kerremans, 1902–1903

Figs **84–99**; 312–316.

Tribe. Nascionini Hołyński, 1988.

General distribution. Australia, New Caledonia, New Zealand. **Australian distribution.** Nineteen Australian species distributed along the east coast, including associated coastal ranges, from North Queensland to Tasmania. The greatest diversity of species is found in North Queensland and northern New South Wales–south-eastern Queensland. General state distributions include *Nascioides bicolor* (NSW), *N. carissimus* (NSW), *N. costatus* (NSW), *N. elderi* (NQld), *N. elessarellus* (NQld), *N. falsomultesimus* (NSW-SQld), *N. macalpinei* (NSW), *N. multesimus* (NSW-SQld), *N. mundus* ((NQld), *N. nulgarra* (NSW-SQld),

N. olliffi (NSW), *N. parryi* (Tas, SA, Vic, ACT, NSW, Qld), *N. pulcher* (NSW--SQld), *N. quadrinotatus* (Tas, Vic), *N. storeyi* (NQld), *N. subcostatus* (NQld), *N. tillyardi* (NSW), *N. viridis* (NSW–SQld), *N. walfordorum* (NQld).

Comments. Individual species range in size from about 5 mm (*Nascioides macalpinei*) to 12 mm (*N. subcostatus*) and are often metallic green (e.g. *Nascioides falsomultesimus*, *N. multesimus*, *N. mundus*) and/or with prominent spots or bands on the elytra (e.g. *N. elessarellus*, *N. olliffi*, *N. parryi*, *N. subcostatus*, *N. tillyardi*). However, *Nascioides costatus* is somewhat cryptic in colouration and thus difficult to see when on standing or fallen tree trunks.

Adult *Nascioides* are mainly encountered from late spring to mid-summer, can be active in sunlight, and usually are found resting on foliage or are attracted to freshly fallen trees and branches. Rarely are they associated with flowers. Habitats include rainforest, open sclerophyll forest and woodland, with *Nascioides parryi* being particular to the latter two. *Nascioides parryi* is a robust elongate species that varies in length from 7–12 mm with distinctly orange and black patterned elytra. It is the most widely recorded member of the genus, usually found at higher elevations, and distributed from south-eastern Queensland, south through eastern New South Wales, Victoria, South Australia and Tasmania (Williams 1987; Cowie 2001; Williams and Bellamy 2002).

A high proportion of species are restricted to rainforest though several, such as *Nascioides pulcher*, extend into adjacent wet sclerophyll forest where appropriate fissured-bark larval food plants occur. Clerid beetles have been recorded attacking emerging adult *Nascioides falsomultesimus* and spiders are also recorded as predators (Williams 1987). Larval food plant records include *Nascioides costatus* and *N. falsomultesimus* from *Ceratopetalum apetalum* (Cunoniaceae), *N. elderi* from *Flindersia brayleyana* (Rutaceae/Flindersiaceae) and *Argyrodendron peralatum* (Malvaceae), *N. viridis* from *Argyrodendron actinophyllum*, *N. falsomultesimus* and *N. multesimus* from *Doryphora sassafras* (Atherospermataceae), *N. falsomultesimus* from *Schizomeria ovata* (Cunoniaceae) and *Sloanea woollsii* (Elaeocarpaceae), *N. parryi* from *Eucalyptus globulus*, *E. tindaliae* (Myrtaceae), *N. pulcher* from *Backhousia myrtifolia*, *B. sciadophora*, *Syzygium floribundum*, *S. smithii* (Myrtaceae), *N. quadrinotatus* from *Nothofagus cunninghamii* (Nothofagaceae), and *N. nulgarra* and *N. tillyardi* from *Nothofagus moorei*.

Adult *Nascioides carissimus*, *N. costatus*, *N. multesimus* and *N. quadrinotatus* are recorded from *Acacia* (Fabaceae) foliage, *N. carissimus* from that of *Angophora* (Myrtaceae), *N. macalpinei* from foliage of *Nothofagus moorei* (Nothofagaceae), *N. parryi* from *Corymbia tessellaris*, *Eucalyptus* spp. and *Leptospermum scoparium* (Myrtaceae), *N. pulcher* from *Syzygium* sp. (Myrtaceae), *N. quadrinotatus* from *Nothofagus cunninghamii* and *N. tillyardi* from *Nothofagus moorei*. Adult *Nascioides olliffi* have frequently been encountered (e.g. Mt Kaputar National Park, NSW) resting on the sunlit branches of *Casuarina* sp. (Casuarinaceae).

Single species are known from New Caledonia (*Nascioides caledonicus*) and New Zealand (*N. enysii*). *Nascioides enysii*, which bears a striking resemblance to the Tasmanian and Victorian species *N. quadrinotatus*, breeds in several species of *Nothofagus* (Dumbleton 1932). However, the larval food plant associations of *N. caledonicus* are unknown.

General references. Bellamy (2002, 2003); Bellamy *et al.* (2013); Burns and Burns (1992); Cowie (2001); Lawrence and Lemann (2019); Matthews (1985); Peterson (1989); Williams (1987, 2002, 2020a); Williams and Bellamy (2002); Williams and Watkins (1985).

Neobubastes Blackburn, 1892

Figs 317, 318.

Tribe. Pterobothrini Volkovitsh, 2001

General distribution. Australia. **Australian distribution.** A small genus of four species largely restricted to Western Australia (*Neobubastes flavovittata*, *N. nickerli*, *N. obscura*.) but with *Neobubastes aureocincta* recorded from the Northern Territory, Queensland, Victoria and South Australia.

Comments. Genus earlier given as *Castelnaudina*, *Castelnaudia* and *Eububastes*.

Species possess a metallic lustre, the colour of *Neobubastes nickerli* blue, that of *N. obscura* bronze–black with a weak blue–violet hue, and *N. aureocincta* and *N. flavovittata* with contrasting lateral margins on the elytra. Depending on species, adults range in size from 8 to 18 mm. *Neobubastes flavovittata* has been collected on *Allocasuarina campestris* (Casuarinaceae) and *Thryptomene/Baeckea* (Myrtaceae), adults of *N. nickerli* have been observed feeding on the foliage of *Acacia* (Fabaceae) and larvae of *Neobubastes obscura* have been recorded from the trunks of *Melaleuca lateriflora* (Myrtaceae)

(Hawkeswood 1987b; Bellamy and Peterson 2000a; Bílý and Powell 2017).

General References. Bellamy (2002, 2003); Bellamy and Peterson (2000a); Bellamy *et al.* 2013; Bílý and Powell (2017); Lawrence and Lemann (2019).

Neobuprestis Kerremans, 1902–1903

Fig. 319.

Tribe. Buprestini Leach, 1815

General distribution. Australia. **Australian distribution.** Three species: *Neobuprestis frenchi* (Vic, NSW), *N. peroni* (SA, Vic, WA) and *N. williamsi* (NSW).

Comments. Genus close to *Zulubuprestis* from southern Africa. Obenberger (1958) erected the genus *Balthasarella* to accommodate the then enigmatic species *Balthasarella melandryoides* (which Obenberger placed in the Polycestinae; discussed in Bellamy 1994) but this proved a synonym of *Neobuprestis frenchi*, itself originally described as *Strigoptera frenchi* (Bellamy 2002). However, *Balthasarella* is retained as a subgenus of *Neobuprestis* (Levey and Bellamy 2013).

The genus consists of moderately sized beetles (14–20 mm), elongate–cubcylindrical in form, and metallic brownish copper to bronze. Adults are generally active December to March. *Neobuprestis williamsi* is recorded from open forest and cool temperate rainforest. The species has been collected resting on *Ozothamnus* foliage at Barrington Tops, and numbers of adults have been observed flying to and resting on broad-leaved *Acacia* foliage (summit of Mt Banda Banda). One specimen was collected in cool temperate rainforest in the Barrington Tops resting on the undersurface of a polypore fungus (possibly *Ganoderma*) (B. Williams pers. obs.). Goudie (1920) commented on *Neobuprestis peroni* (as *Strigoptera australis*): 'Twenty years ago it used to occur in considerable numbers in the Birchup district (Vic), appearing about the middle of January, when it would be found in the mornings clinging to the wheat stubble, low mallee shoots, &c.'

Selected collection localities: *Neobuprestis peroni* (Kangaroo Island/Mindarie [SA]), Birchip/Sea Lake [Vic], Holt Rock/Hurlstone [WA]); *N. frenchi* (Lake Mtn/Ben Cairn [Vic]), Kosciuszko National Park [NSW]); *N. williamsi* (Barrington Tops/Werrikimbe/New England/Border Ranges National Parks [NSW]).

General references. Bellamy (2002, 2003); Burns and Burns (1992); Lawrence and Lemann (2019); Levey and

Bellamy (2013); Matthews (1985); Obenberger (1958); Williams (2002, 2020a).

Neocuris Saunders, 1868

Figs **100–105.**

Tribe. Curidini Hołyński, 1988

General distribution. Australia. **Australian distribution.** About 30 described species, the genus being recorded from all states except Tasmania. There are several undescribed *Neocuris* species, but the genus has not been revised since Carter (1928a). Only a few species are recorded from both eastern states and Western Australia, *Neocuris discoflava*, *N. thoracica* and *N. viridimicans* being examples of species with mainland-wide distributions.

Comments. Species of *Neocuris* generally resemble *Anthaxia*, a speciose genus absent from Australia but with an otherwise cosmopolitan distribution. Members of *Neocuris* are small beetles that range in size from barely 2 mm (*Neocuris anthaxoides*) to about 6 mm (*N. crassus*, *N. discoflava*), and occur in a variety of vegetation types including heath, shrub complexes, woodland, eucalypt-dominated sclerophyll forest, and rainforest. Species are generally metallic in lustre, often with a contrasting reflected colour on the lateral margin of the pronotum. *Neocuris carnabyae*, *N. guerinii* and *N. discoflava* possess conspicuous yellow colouration on the disc of the elytra. Most species are glabrous but a number (e.g. *N. coerulans*, *N. pubescens*) are clothed in moderately dense erect setae.

Although individuals are occasionally collected on foliage, they are mainly to be found associated with flowers, especially mass-flowering trees and shrubs with open and shallow floral structures. Records for adult *Neocuris* associated with flowers (generally feeding on nectar, to which their mouthparts are adapted), primarily are those for Myrtaceae, although this may reflect a bias towards collecting from myrtaceous species. Records for *Neocuris* from Myrtaceae include *Neocuris anthaxoides* (*Leptospermum polygalifolium*); *N. asperipennis* (*L. polygalifolium*); *Neocuris coerulans* (*Leptospermum polygalifolium*, *Syzygium smithii*); and *N. cuprilatera* (*Baeckea* sp., *Eucalyptus* spp., *Leptospermum polygalifolium*). However, adult feeding records often include *Neocuris* species that are known from a suite of plant families, such as *Neocuris gracilis* (*Angophora woodsiana*, *Backhousia myrtifolia*, *Baeckea stenophylla*, *Kunzea ambigua*, *Leptospermum juniperinum*, *L. polygalifolium*, *L. trinervium*, *L. whitei* [Myrtaceae], *Actinotus helianthi* [Apiaceae], *Epacris microphylla* [Epacridaceae], *Bursaria*

spinosa [Pittosporaceae], and *Cassinia* sp. [Asteraceae]); *Neocuris guerinii* (*Eucalyptus* spp., *Kunzea ambigua*, *Kunzea* spp., *Leptospermum polygalifolium* and *Leptospermum* spp. [Myrtaceae], *Actinotus helianthi* [Apiaceae], *Conospermum* sp. [Proteaceae]); *Neocuris viridimicans* (*Eucalyptus* spp. [Myrtaceae], *Helichrysum* sp. [Asteraceae], and *Nuytsia floribunda* [Loranthaceae] (Fig. 571) – *Nuytsia floribunda* is also recorded as a nectar source for *Neocuris discoflava*). An undescribed *Neocuris* occurring in lowland rainforest in New South Wales is recorded from *Euroschinus falcatus* (Anacardiaceae). Larvae of *Neocuris gracilis* are known from *Pultenaea villosa* (Fabaceae).

General references. Bellamy (2002, 2003); Bellamy *et al.* (2013); Carter (1928a); Lawrence and Lemann (2019); Volkovitsh and Hawkeswood (1987).

Notobubastes Carter, 1924

Fig. **106.**

Tribe. Bubastini?

General distribution. Australia. **Australian distribution.** Consists of four species. Two of these, *Notobubastes aurosulcata* and *N. occidentalis*, are restricted to Western Australia, *N. costatus* is recorded from Western Australia and South Australia, and *N. orientalis* is widely recorded in Queensland.

Comments. The genus is closely related to *Microcastalia*. Species are somewhat cylindrical–elongate in shape with a metallic lustre. Habitats include woodland. Adults are often associated with foliage. *Notobubastes costatus* has been reared from billets of *Exocarpus aphyllus* (Santalaceae) (M. Powell, cited in Lang 2020). E. Adams (unpubl. records) cites adult *Notobubastes orientalis* collected in Queensland from 'quinine bush' (*Petalostigma pubescens* – Picrodendraceae, a small tree occurring in Queensland, Northern Territory, New South Wales and Western Australia) during February.

General references. Bellamy (2002, 2003); Bellamy and Peterson (2000a); Carter (1924, 1928a); Hołyński (1988); Lawrence and Lemann (2019); Matthews (1985).

Notographus Thomson, 1879

Fig. **107.**

Tribe. Curidini Hołyński, 1988

General distribution. Australia. **Australian distribution.** Four species: *Notographus hieroglyphicus* (Qld), *N. sulcipennis* (Qld), *N. uniformis* (Vic, SA, WA) and *N. yorkensis* (Qld, WA).

Comments. These are small species (approx. 4–9 mm), and are largely bronze–brown. The elytra of *Notographus sulcipennis* possess distinctive longitudinally indented sculpturing, an attribute somewhat present in *Australorhipis aphanochila* (Fig. 38), but the elytra are generally smooth or only indistinctly sculptured in other *Notographus* species. In colouration and general form *Notographus* may be confused with *Anthaxoschema* (Figs 35, 36) and *Anilara* (e.g. Figs 31, 297), but differ from these in having strongly convergent eyes behind and the lateral margins of the pronotum more rounded and distinctly narrower than the base of the elytra.

Adults are associated with foliage and branch stems during the day and have been recorded from 'mallee' woodland and rainforest (Matthews 1985; G. Williams pers. obs.). Individuals have been reared from dead branches of an *Acacia* sp. (Fabaceae) collected in woodland (vic. Mt Kaputar National Park, CNSW) (G. Williams unpubl. records).

General references. Bellamy (2002, 2003); Bellamy and Peterson (2000a); Lang (2020); Lawrence and Lemann (2019), Matthews (1985).

Pseudanilara Théry, 1911
Fig. **108**.

Tribe. Curidini Hołyński, 1988

General distribution. Australia. **Australian distribution.** Eleven species, the genus widely recorded from mainland Australia (10 species [Bellamy 2002]) as well as Tasmania (4 species [Cowie 2001]); with *Pseudanilara kerremansi* apparently restricted to Tasmania (Bellamy 2002).

Comments. Adults are metallic-hued, usually of a single colour (e.g. bronze–black, greenish purple) or with contrasting bright reflections, these usually on the pronotum (*Pseudanilara pilosa*). Individual species range in size from approximately 3 mm (*Pseudanilara purpureicollis*) to 10 mm (*P. cupripes*). But there can be considerable variation in size, for example *P. cupripes* ranging from 4 to 10 mm or so. The genus occurs in woodland, mallee, eucalypt-dominated wet and dry sclerophyll forest formations and rainforest, with individuals generally active particularly from October to February but with some records either side of these months. Adults are collected by sweeping or beating foliage and branches but have occasionally been recorded from flowers. Known plant association records include adult *Pseudanilara cupripes* from flowering *Corymbia tessellaris* and *Syzygium floribundum* (Myrtaceae) and the larval hosts *Ceratopetalum gummiferum*, *Schizomeria ovata* (Cunoniaceae), *Backhousia myrtifolia*, *B. sciadophora*, *Eucalyptus crebra*, *E. paniculata*, *Kunzea ambigua*, *Rhodomyrtus psidioides*, *Syzygium floribundum* (Myrtaceae); *P. piliventris* on foliage of dying *Eucalyptus* spp.; and adult *P. purpureicollis* from *Casuarina cunninghamiana* (Casuarinaceae) and on the foliage of dying *Eucalyptus* spp. (Myrtaceae). *Pseudanilara cupripes*, a species seasonally common in wet sclerophyll forest and lowland rainforests, has occasionally been observed at lights (Williams 1982).

A number of species were earlier placed in *Neotorresita*, *Anthaxia*, *Neocuris*, *Melobasis* and *Melanophila*. There was particular historical controversy concerning *Pseudanilara*, described in 1910 by André Théry, and *Neotorresita*, which was described by Jan Obenberger in 1923; the validity of the former and the synonymous status of the latter was the subject of claims and rebuttals in a series of papers over many years by Obenberger and H. J. Carter (see Bellamy and Peterson (2000a)).

General references. Bellamy (2002, 2003); Bellamy and Peterson (2000a); Bellamy *et al.* (2013); Burns and Burns (1992); Carter (1926); Cowie (2001); Hawkeswood (2007d); Lawrence and Lemann (2019); Matthews (1985); Williams (1985, 2002, 2020a); Williams and Williams (1983).

Selagis Dejean, 1836
Figs **109–114**; 320–323.

Tribe. Curidini Hołyński, 1988

General distribution. Australia and the Neotropics (Chile) (Bellamy 2003; also see discussion in Bílý and Volkovitsh 2007). **Australian distribution.** Twenty-four species, the genus being widely distributed in mainland Australia but apparently absent from Tasmania.

Comments. Australian species were previously placed in *Curis*, with *Selagis splendens*, a species with distinctively shortened elytra, earlier placed in *Neocuropsis* (Bellamy 2002). Individuals are generally metallic in colour, commonly bright green, bluish purple or coppery bronze (e.g. *Selagis corusca*, *S. viridicyanea*) and frequently with contrasting reflective hues on the pronotum and/or elytra (e.g. *S. adamsi*, *S. caloptera*, *S. yalgoensis*) (Carter 1928a). *Selagis atrocyanea*, however, is black. In *Selagis caloptera* there is a gradation of colouration, with examples (= variety *formosa* Gestro) from Queensland and inland New South Wales (e.g. Mt Kaputar National Park) exhibiting a pronounced brilliance of metallic hue. Species range in approximate size from about 8 mm

(*Selagis splendens*) to about 15 mm (*S. yalgoensis*); however, individuals within species may vary considerably in length (as in *Selagis aurifera*, *S. splendens*). The genus has been collected in a diversity of shrubland, woodland, sclerophyll forest and rainforest habitats, in particular where Myrtaceae abound. Several species, for example *Selagis caloptera* and *S. aurifera*, are occasionally present on the same flowering plant. Adults may feed on nectar and are usually associated with flowering plants that possess open, shallow floral structures (rather than tubular corollas), as in Myrtaceae and many rainforest taxa. The mouthparts are elongate, this character facilitating feeding on floral nectar (figured in Obenberger 1955). Individual species are potential pollinators.

Most host plant records are for adults. *Selagis aurifera* has been recorded on flowers of *Angophora floribunda*, *A. hispida*, *Melaleuca armillaris* and *Syzygium floribundum* (Myrtaceae), *S. caloptera* on flowers of *Angophora floribunda*, *A. hispida*, *Baeckea virgata*, *Eucalyptus* spp., *Kunzea ericoides*, *Leptospermum polygalifolium* (Myrtaceae) and *Bursaria spinosa* (Pittosporaceae), *S. splendens* on *Baeckea virgata*, *Syzygium floribundum*, *S. smithii* (Myrtaceae) and *Bursaria spinosa* (Pittosporaceae), and *S. viridicyanea* on *Corymbia gummifera*, *Eucalyptus acmenioides*, *E. moluccana* and *E. ochrophloia* (Myrtaceae).

Larval records include *Selagis aurifera* from *Eucalyptus gracilis* (Myrtaceae), *S. caloptera* from *Allocasuarina littoralis* (Casuarinaceae) and *Acacia* sp. (Fabaceae), and *S. intercribrata* from a species of Casuarinaceae (Hawkeswood 1988).

General references. Bellamy (2002, 2003); Bellamy *et al.* (2013); Bílý and Volkovitsh (2007); Carter (1928a); Cowie (2001); Hawkeswood (1978); Lawrence and Lemann (2019); Matthews (1985); Obenberger (1955); Williams (1977, 2002, 2020a); Williams and Adam (1998, 2019); Williams and Williams (1983).

Theryaxia Carter, 1928
Fig. **115**.
Tribe. Curidini Hołyński, 1988
General distribution. Australia. **Australian distribution.** One species (*Theryaxia suttoni*), recorded from Queensland and New South Wales.

Comments. *Theryaxia suttoni* is a small species (4–5 mm in length), with the head and pronotum distinctly bronze, and the elytra metallic green with bronze margins. The head is noticeably wider than the anterior margin of the pronotum. The species inhabits sclerophyll woodland and has been recorded from several localites in central New South Wales (e.g. Pilliga East State Forest, Moonbi Hill vic. Tamworth) and south-eastern Queensland (Chinchilla, vic. Milmerran, Stanthorpe). Adults are associated with the foliage of *Callitris* (Cupressaceae) pines, and larvae have been reared from billets of *Callitris columellaris* (Hawkeswood 1986b) and *C. glaucophylla* (Webb *et al.* 1988). Given the broad distribution of the host genus *Callitris* the geographical range inhabited by *Theryaxia suttoni* is likely to be greater than the few known collection localities suggest.

General references. Bellamy (2002, 2003); Bellamy *et al.* (2013); Carter (1928b); Hawkeswood (1986b, 2007e); Lawrence and Lemann (2019); Webb *et al.* (1988).

Torresita Gemminger & Harold, 1869
Figs **116–118**; 324.
Tribe. Curidini Hołyński, 1988
General distribution. Australia. **Australian distribution.** The Australian fauna consists of two described species but at least one undescribed species is also known (G. Williams unpubl. records). *Torresita cuprifera* is widely recorded from eastern New South Wales, but also occurs in Queensland and Victoria. *Torresita parallela* is recorded from New South Wales and Queensland. *Torresita* may also occur on Lord Howe Island (C. Reid pers. comm.).

Comments. Adults are metallic bronze to dark greenish bronze and are approximately 6–13 mm in length. Species occur in woodland, wet and dry sclerophyll forest, and rainforest. Adult *Torresita cuprifera* are very active in sunlight, feed on nectar, and generally frequent flowers with shallow, open floral structures. The species has been recorded from the flowers of *Angophora hispida*, *Baeckea virgata*, *Eucalyptus sieberi*, *Kunzea ericoides*, *Leptospermum polygalifolium*, *L. whitei*, *Melaleuca* sp., *Syzygium floribundum* (Myrtaceae), *Ceratopetalum gummiferum* (Cunoniaceae), and *Alphitonia excelsa* (Rhamnaceae). Adults have been reared from the threatened myrtaceous tree *Rhodomyrtus psidioides* (Williams 1985, 2018), plants having been subsequently killed *en masse* throughout their range by the South American fungal pathogen 'myrtle rust' (*Austropuccinia psidii*), and generally only surviving as root suckers constantly impacted by pathogen spores (Williams 2018). *Torresita cuprifera* has occasionally been observed at lights (Williams 1982).

General references. Bellamy (2002, 2003); Bellamy *et al.* (2013); Hawkeswood (2007f); Lawrence and Lemann

(2019); Williams (1985, 2021); Williams and Adam (1998, 2019).

SUBFAMILY STIGMODERINAE SAUNDERS, 1871

As considered here Stigmoderinae comprises a single nominate tribe, Stigmoderini.

Calodema Gory & Laporte, 1838

Figs **119–121**; 325–327.

Tribe. Stigmoderini Lacordaire, 1857

    General distribution. Australia, New Guinea, eastern Indonesia. The genus is well represented in New Guinea, with about 13 species occurring there (Nylander 2008; Pineda and Curletti 2020). **Australian distribution.** Four species confined to eastern coastal Queensland and New South Wales south to about Bulahdelah: *Calodema regale* (NQld-NNSW), *C. wallacei* (NQld, New Guinea), *Calodema rubrimarginata* (NQld), *C. plebeium* (NQld).

    Comments. These are large attractively coloured beetles, with species extending to greater than 45 mm in length. There are degrees of variation in the extent of individual areas of colour and pattern in some species, such as as in *Calodema regale* in which the yellow elytral colour and surface punctuation, size of the large red spots on the pronotum, and the number and size of the spots on the ventral surface of the abdomen exhibit distinctive regional forms. Overall pigmentation may also vary, for example in *Calodema plebeium* in which adults may display red, pink or orange-like colour themes.

    Adults of all species are diurnal and mostly encountered from mid- to late summer. The northern Queensland species *Calodema plebeium*, for example, is active from December to February with populations of the more widespread *C. regale* active over the same period. Species inhabit tropical and subtropical rainforest, but will exhibit 'hill-topping' behaviour, there feeding on nectar from low exposed shrubs, such as *Leptospermum* and *Kunzea*, flowering on rocky escarpments and mountaintops. Though large beetles, adults can be surprisingly well-hidden from view among blossoms when feeding. Individual *Calodema regale* observed in North Queensland and northern New South Wales opened their wings and flew off when released from height, rather than tumbling to the ground and feigning death.

    Mouthparts are prolonged and adapted for feeding on nectar, in the process accumulating relatively large pollen loads on the underbody before moving to adjacent co-flowering plants. Consequently adults are potentially important agents of long distance pollen transfer in rainforests (Williams 2020a, 2021). Adult host plant records include *Calodema plebeium* on flowers of *Corymbia gummifera*, *C. intermedia*, *Syzygium* sp. (Myrtaceae), *Buckinghamia celcissima*, *Macadamia whelanii* (Proteaceae) and *Melicope elleryana* (Rutaceae); *Calodema regale* on *Elaeocarpus angustifolius* (Elaeocarpaceae), *Bauhinia monandra* (Fabaceae), *Corymbia gummifera*, *C. intermedia*, *Eucalyptus moluccana*, *Kunzea* sp., *Leptospermum* sp. *Syzygium* spp., *Tristaniopsis laurina* (Myrtaceae), *Buckinghamia celcissima*, *Macadamia whelanii* (Proteaceae), *Cuttsia viburnea* (Rousseaceae), *Melicope elleryana* and *M. micrococca* (Rutaceae), and *Calodema rubrimarginata* on flowers of *Melicope elleryana*.

    Selected collection localities: *Calodema regale* (Coorabakh National Park, Dorrigo, Acacia Plateau [NSW], Mt Glorious, Innisfail, Kuranda [Qld]); *C. wallacei* (Cape York, [NQld], Papua [Indonesia], Papua New Guinea); *C. rubrimarginata* (Cape York [Qld]); *C. plebeium* (Cairns, Garradunga, Cape York [NQld]).

    General references. Barker (1993b); Bellamy (2002, 2003); Bellamy et al. (2013); Gardner (1989); Lawrence and Lemann (2019); Nylander (2008); Williams (2002, 2020a); Williams and Adam (2010).

Calotemognatha Peterson, 1991

Figs **122, 123**.

Tribe. Stigmoderini Lacordaire, 1857

    General distribution. Australia. **Australian distribution.** Three species and several subspecies recorded from Western Australia and Victoria: *Calotemognatha laevicollis* (WA), *C. varicollis* (WA) and *C. yarelli* (Vic, WA). Records mostly concentrated in south-western Western Australia.

    Comments. Species of *Calotemognatha* were previously placed in *Temognatha*. Features of the ovipositor separate *Calotemognatha* from *Temognatha* (Peterson 1991). These are moderately sized beetles associated with 'mallee' woodland and low open forest. Beetles feed on nectar and are particularly associated with Myrtaceae. Adult host records include flowers of *Nuytsia floribunda* (Loranthaceae), *Eucalyptus foecunda*, *E. uncinata*, *Melaleuca acuminata* and *M. thymifolia* (Myrtaceae). As with other larger-sized Stigmoderini adult *Calotemognatha* are potentially important contributors to out-crossing of self-incompatible plant species.

Selected collection localities: *Calotemognatha yarelli* (vic. Hyden, Moore River district, Swan River, WA), *C. varicollis* (Wanneroo, WA).

General references. Bellamy (2002, 2003); Bellamy and Peterson (2000a); Gardner (1989); Lawrence and Lemann (2019); Peterson (1991).

Castiarina Gory & Laporte, 1838

Figs **124–171**; 328–465, 469.

Tribe. Stigmoderini Lacordaire, 1857

General distribution. Largely found in Australia; however, several species are known from New Guinea (e.g. *Castiarina borealis*, *C. dryadula*, *C. loriae*, *C. meeki*, *C. shelleybarkeri*). **Australian distribution.** The Australian fauna comprises about 470 species, the genus being widely distributed in mainland Australia and Tasmania. *Castiarina loriae* occurs in New Guinea with additional records from the Northern Territory (Darwin) and Queensland (Bowen, Rockhampton).

Comments. Species of *Castiarina* are small elongate beetles that range in size from about 5 mm (e.g. *C. ariel*, *C. jackhasenpuschi*, *C. puella*) to approximately 20 mm or more in length (*C. bremei*, *C. rollei*, *C. scintillata*) and are to be found in heathland, open and closed forest, shrub communities and woodland; they constitute an iconic element of the Australian buprestid and beetle biota more generally. Despite historically extensive collecting new species continue to be discovered (e.g. *Castiarina notocrux* from southern Western Australia, *C. testudocaput* from Turtle Head Island at the tip of Cape York Peninsula [Hutchinson and Allsopp 2022]). Adults are mostly active early spring to early autumn, though individual species may emerge within more seasonally circumscribed periods, with mid-spring to mid-summer generally being the time when most species can be encountered. In montane zones peak occurrences may occur through December and January. Seasonal rainfall patterns are thought to strongly influence emergence of adults, with the emergence of many punctuated by years of seeming absence. Shade and cool temperature also influence daily patterns of activity.

These species, despite great variation in size, generally exhibit an elongate body plan and moderately long serrate antennae. However, there are instances of departure from the overall uniformity of form. The males of the Papua New Guinean *Castiarina shelleybarkeri* possess distinctive bipectinate/biflabellate antennae. This is a character also exhibited by males of *Castiarina variegata* (syn. *Hypostigmodera variegata*), a species occurring in northern New South Wales in cool temperate, subtropical and littoral rainforests (Bellamy and Nylander 2007). Blackburn (1892), describing *Hypostigmodera variegata* in the *Transactions of the Royal Society of South Australia*, writes: 'The extraordinary insect on which I found this genus seems to be a perfectly typical *Stigmodera* in all respects except that the antennae of the male are strongly biflabellate, resembling very closely those of the male in the *Elaterid* genus *Euphemus*; the antennae of the female are those of an ordinary *Stigmodera*' (*Castiarina* at that time being a subgenus of *Stigmodera*). In the relatively large black and yellow *Castiarina bremei* the tarsomeres are conspicuously enlarged and the tarsal pads (pulvilli) greatly expanded; however, the purpose of such adaptations is not understood.

The majority of species possess patterned, often brightly coloured elytra; these, as earlier discussed, frequently suggest a warning (aposmatic) function so as to deter predators. Unfortunately the vibrant colour of many species in life darkens in museum specimens. Although numerous species (e.g. *Castiarina decemmaculata*, *C. violacea*, *C. sexplagiata*, *C. skusei*) are widespread within parts of the mainland, with some shared with Tasmania (e.g. *C. australasiae*, *C. erythromelas*, *C. rufipennis*, *C. undulata*), the number of species distributed between eastern and western Australia is relatively few (e.g. *C. bimaculata*, *C. castelnaudi*, *C. cyanipes*, *C. pallidiventris*). In southern mainland Australia a broadly defined western Victoria–South Australia–south-west Western Australian fauna can be distinguished from that of the south-east of the Australian continent. Equally, there are species confined to, and distinctive of north-eastern Queensland as well as south-eastern Queensland and north-eastern New South Wales (Barker 2006a); these two areas are biodiversity 'hot spots' for many mesic-adapted plants and animals. In addition, there is an apparent latitudinal sifting of species (a phenomenon found in other groups of Coleoptera such as Carabidae), this running north and south along the eastern seaboard and associated near-coastal ranges and tablelands. There are also species distinctive of higher altitudes (e.g. *Castiarina attenuata*, *C. eborica*, *C. helmsi*, *C. macquillani*, *C. zecki*) and not surprisingly there are numerous species characteristic of rainforest, the presence of these in the field veiled by the height of the tree canopies that they frequent. However, despite the historical enthusiasm among collectors of the genus, some species remain known by single or few specimens only (e.g. *Castiarina tropica* from Cape York Peninsula [Qld]; the unique holotype of *C. atrocoerulea* from unknown locality [Barker 2006a]) and there are almost no published records based on long-term surveys that indicate the diversity and distribution of species at regional or local scales.

Castiarina are nectar feeders and can be seasonally common, particularly on flowering Myrtaceae, this association suggesting a co-evolutionary relationship. But the range of flowering plants that adult *Castiarina* have been recorded from is extensive (Bellamy *et al.* 2013), their presence strongly influenced by flower structure (typically open and shallow 'bowl-shaped' [see discussion in Williams 2020a, 2021; Williams and Adam 2010]) and the presence of nectar, and possibly pollen, as a reward. There is little evidence, however, to suggest instances of fidelity of adult *Castiarina* to particular flowering plants, the association being that of temporary attraction to available floral food resources, with beetles 'switching' to whatever flowering plants may next be available, or that offer greater recruitment cues such as scent plumes or the visual conspicuousness of available blossoms contrasting with foliage. And although most records are of adults visiting white flowers this is primarily owing to the general prevalence of white-flowered plants in the landscape, not necessarily a preference for that colour. Many, if not all, adult *Castiarina* can be considered pollinators of numerous mass-flowering trees and shrubs. But there are examples of *Castiarina* being visitors to, and pollinators of, flowering plants that depart from the myrtaceous theme of structure or white colour. For example, Hawkeswood (2007c) cites *Castiarina crenata* and *C. sexplagiata* as pollinators of *Conospermum taxifolium* (Proteaceae), and Williams (2020b) lists *C. neglecta* as the solitary buprestid pollinator within a taxonomically diverse guild of more than 170 insect species that visit the dull orange–brown flowers of the mangrove *Avicennia marina* (Acanthaceae) in northern New South Wales. Although adult flower visitation records are many, larval host records are significantly fewer: *Castiarina decipiens* (*Clerodendrum floribundum* [Lamiaceae]), *C. insularis* (*Eucalyptus* sp. [Myrtaceae]), *C. praetermissa* (*Eucalyptus microcorys* [Myrtaceae]), *C. parallela* (*Dodonaea* sp. [Sapindaceae]), *C. producta* (*Muellerina eucalyptoides* [Loranthaceae]), *C. rufipennis* (*Acacia cyclops, A. ulicifolia* [Fabaceae]) and *C. verdiceps* (*Acacia* sp. [Fabaceae]). Barker (1979) notes the bodies of dead adult *Castiarina uptoni* were found in tubular highway marker posts north of Barrow Creek,

Northern Territory. He considered that they had been placed there by predaceous insects. Although live adults were subsequently collected in the same area their food plants were not recorded. However, in Western Australia larvae and adults of *Castiarina uptoni* were found associated with the low-growing shrub *Dicrastylis georgei* (Lamiaceae) (Barker 1993a). Adults have been found on the leaves, with larvae tunnelling and pupating in live stems.

General references. Barker (1986, 2006a); Bellamy (2002, 2003); Bellamy *et al.* (2013); Burns and Burns (1992); Carnaby (1987); Carter (1916, 1931); Cowie (2001); Gardner (1989); Lawrence and Lemann (2019); Matthews (1985); Williams (2002, 2020a, 2021); Williams and Adam (1998, 2010, 2019).

Metaxymorpha Parry, 1848

Figs **175–179**; 466.

Tribe. Stigmoderini Lacordaire, 1857.

General distribution. Australia, eastern Indonesia, and is especially speciose in New Guinea (Nylander 2008). **Australian distribution.** Five species are known from Australia (*Metaxymorpha gloriosa, M. hauseri* [NQld], *M. hanloni* [NT], *M. grayii, M. imitator* [NNSW]), the genus distributed from northern Australia, and eastern Australia to about the mid-north coast of New South Wales.

Comments. These are relatively large buprestids (~25–30 mm). Adults are found on blossoms of a wide variety of trees, including species with large, shallow flowers (as in *Corymbia, Leptospermum*) and those in which blossoms are somewhat tubular, or have a degree of depth effect (e.g. *Elaeocarpus, Melicope*). Adult hosts include families in which flowers are actinomorphic (Myrtaceae) or zygomorphic (Proteaceae). Mouthparts are elongated, and adapted to feeding on nectar. Species are associated with rainforest, or flowering rainforest species growing in or on the margins of eucalypt-dominated wet sclerophyll forest. Adults of the southern Australian species *Metaxymorpha grayii* and *M. imitator* are active in summer. Species inhabit lowland and sub-montane near-coastal rainforest as well as hinterland montane rainforest.

Host plant records for *Metaxymorpha gloriosa* are extensive. Adults have been recorded on flowers of *Aphananthe philippinensis* (Cannabaceae), *Elaeocarpus angustifolius* (Elaeocarpaceae), *Corymbia gummifera, C. intermedia, Eucalyptus moluccana, Eucalyptus* spp.,

Leptospermum spp., *Syzygium tierneyanum*, *S. kuranda* (Myrtaceae), *Buckinghamia celcissima*, *Grevillea baileyana*, *Macadamia whelanii* (Proteaceae), *Psydrax odorata* (Rubiaceae), *Melicope elleryana* (Rutaceae) and *Synima cordieri* (Sapindaceae) (Volkovitish *et al.* 2003). Larvae have been found in *Casuarina cunninghamiana* (Casuarinaceae), *Cyclophyllum multiflorum* (Rubiaceae), *Guioa lasioneura* and *Sarcopteryx reticulata* (Sapindaceae). Volkovitish *et al.* (2003) described the behaviour of larvae of *Metaxymorpha gloriosa* in *Guioa lasioneura* (Sapindaceae). After emerging from eggs laid on the upper part of the trunk, larvae bore nearly straight tunnels into the heartwood and then proceed to the basal sections of the tree and its roots. In the base of the host the mature larvae graze back and forth in vertical open galleries. Larvae construct perpendicular side tunnels that allow ventilation and which are terminated by small holes in the bark, through which frass is expelled. The expelled frass allows recognition of larval activity within the host tree. The authors considered that development of larvae may take several years, noting also that heavily infested trees usually die.

Adult *Metaxymorpha grayii* are recorded from flowers of *Tristaniopsis laurina*, *Baeckea* sp. (Myrtaceae) and *Melicope micrococca* (Rutaceae). Adult host records for *Metaxymorpha hauseri* include *Corymbia gummifera*, *C. intermedia*, *Syzygium* spp., *Tristaniopsis laurina* (Myrtaceae), *Elaeocarpus angustifolius* (Elaeocarpaceae)], *Corymbia intermedia*, *Buckinghamia celcissima*, *Macadamia whelanii* (Proteaceae), *Melicope elleryana* and *M. platynema* (Rutaceae). *Metaxymorpha imitator* is recorded from flowering *Eucalyptus* species.

Metaxymorpha grayii and *M. imitator* are very similar in colour (black and blood red), the resemblance making them difficult to differentiate when feeding on blossoms. The two species appear to be part of a larger mimic complex that includes similarly coloured species of *Temognatha*.

Metaxymorpha hanloni was described by Nylander (2008) based on a single poorly labelled and damaged female specimen 'found dead in a pool after flood' at the Adelaide River, Northern Territory. Twenty-nine millimetres in length, this enigmatic species has a shiny, coppery red pronotum, the elytra reddish brown with a narrow black basal fascia, a broad black postmedial fascia that extends from the suture almost to the margins and the apices black with three spines. The ventral surface, including legs, is metallic green.

Selected locality records: *Metaxymorpha gloriosa* (Cape Tribulation/Mossman/Kuranda [Qld]); *M. grayii* (Lansdowne Escarpment/Acacia Plateau/Legume [NSW]; Mt Glorious [Qld]); *M. imitator* (Manning Valley [NSW]); *M. hanloni* (Adelaide River [NT]); *M. hauseri* (Innisfail/Cairns/Kuranda [Qld]).

General references. Bellamy (2002, 2003); Bellamy *et al.* (2013); Carnaby (1987); Gardner (1989); Lawrence and Lemann (2019); Nylander (2008); Volkovitsh *et al.* (2003); Williams (2002, 2020a, 2021); Williams and Adam (1998).

Stigmodera Eschscholtz, 1829
Figs **180–185**; 467–469.
Tribe. Stigmoderini Lacordaire, 1857.

General distribution. Australia. **Australian distribution.** Seven species, the genus occurring in all mainland states but absent from Tasmania: *Stigmodera cancellata*, *S. gratiosa*, *S. roei* (WA), *S. jacquinotii*, (Qld, NSW, Vic), *S. macularia* (Qld, NSW, Vic, SA), *S. porosa* (Qld) and *S. sanguinosa* (Vic, SA, WA).

Comments. Species are robust and elongate–ovoid in shape, and range in size from about 14 mm (*Stigmodera gratiosa*) to >30 mm (*S. cancellata*). However, individuals can vary considerably in length (e.g. *S. cancellata* – 22–33 mm; *S. macularia* – 20–30 mm). Colour varies from the whole body being brilliant metallic green to green–gold (*Stigmodera gratiosa*), elytra metallic green with areas of red (*S. cancellata*, *S. roei*), red–brownish elytra with black head and pronotum (*S. porosa*, *S. sanguinosa*), and yellow elytra with black punctuation contrasted with black head and pronotum (*S. jacquinotii*, *S. macularia*). *Stigmodera jacquinotii* and *S. macularia* are identical in general appearance, suggesting a mimic complex, a conspicuous distinguishing feature being the elytral apices which are pointed in *S. jacquinotii* but rounded in *S. macularia*. *Stigmodera jacquinotii* (Fig. 468) and *S. macularia* (Fig. 469) possess some resemblance to the rare co-occurring *Temognatha goryi* (Fig. 194), which might indicate a wider mimicry assemblage.

Adult *Stigmodera* can be encountered in woodland, heath and open sclerophyll forest, with most species emerging in spring and being active throughout summer. The Western Australian *Stigmodera gratiosa* has been collected as early as July. Adults feed on nectar and as with the related stigmoderine genera *Castiarina*, *Temognatha* and *Calotemognatha* there is an association between *Stigmodera* and members of the Myrtaceae. Host plant records (for both adults and larvae) almost exclusively reflect this association and include *Stigmodera cancellata* larvae in *Agonis flexuosa* and adults on flowers of *Chamelaucium uncinatum* and *Leptospermum* spp. (Myrtaceae); *Stigmodera gratiosa* larvae in *Agonis flexuosa*, *Leptospermum* spp., *Melaleuca* spp.; and adults on flowers or foliage of *Agonis* spp., *Eucalyptus* spp., *Leptospermum* spp., *Melaleuca* spp. (Myrtaceae) and *Hakea* spp. (Proteaceae); *Stigmodera jacquinotii* adults on flowers of *Angophora hispida* and *Leptospermum polygalifolium* (Myrtaceae); *Stigmodera macularia* adults on flowers of myrtaceous *Angophora bakeri*, *A. hispida*, *Baeckea stenophylla*, *Eucalyptus* spp., *Kunzea ambigua*, *K. ericoides*, *Leptospermum polygalifolium*, *L. trinervium*, *Leptospermum* spp. and *Melaleuca linariifolia*; *Stigmodera roei* larvae in *Agonis flexuosa* and adults on *Chamelaucium uncinatum*, *Leptospermum* spp. and *Melaleuca* spp.; and *S. sanguinosa* adults on flowers of *Nuytsia floribunda* (Loranthaceae), *Leptospermum* spp. and *Melaleuca* spp.

General references. Bellamy (2002, 2003); Bellamy *et al.* (2013); Burns and Burns (1992); Carter (1916); Cowie (2001); Gardner (1989); Hawkeswood (1978); Lawrence and Lemann (2019); Macqueen (1948); Matthews (1985); Nikitin (1979); Williams (2002, 2020a); Williams and Williams (1983).

Temognatha Solier, 1933

Figs **186–216**; 470–502.

Tribe. Stigmoderini Lacordaire, 1857.

General distribution. Australia. **Australian distribution.** Bellamy (2003) cites 83 species; however, a number are poorly known. *Temognatha* occurs throughout the mainland, the Western Australian fauna being particularly rich (37+ spp.), but with only one species (*Temognatha mitchellii*) occurring in Tasmania (Cowie 2001; Bellamy 2002).

Comments. Adults are robust and broadly elongate in form, range in size from about 14 mm (*Temognatha secularis*) to >5.5 cm (*T. heros*), and are diverse in colour and pattern. There can be considerable variation in elytral patterns exhibited by some species. For example, individuals of *Temognatha variabilis* can be found with elytra that are fully black or ochre, with black subfascia, or with distinctive black lateral fascia. In similar fashion the elytra of *Temognatha stevensii* vary from forms devoid of fascia to those in which subfascia are conspicuous. Additional examples include *Temognatha obesissima*, *T. regia*, *T. nickerli* and *T. maculiventris*. Many species of *Temognatha* are to be found in eucalypt-dominated sclerophyll forest and woodland, and are seasonally very diverse in Western Australian mallee but less frequent in rainforest. Adults are nectar feeders; however, the emergence of individual species can be punctuated by several years or more of absence; periods of 8 years or so between emergence events are common. This may be related to rainfall and temperature patterns. Individuals often undertake relatively long-distance flights between co-flowering plants and in so doing can contribute to out-crossing in plant populations. Adults, when present, are most commonly encountered in mid- to late summer, frequently during the hottest months of the year. In eastern New South Wales species (e.g. *Temognatha gemmelli*) are known to 'hill top', occurring on flowering shrubs and trees in elevated situations such as volcanic plugs and escarpments.

Adult host records are numerous (see Bellamy 2013) and include flowering myrtaceous trees and shrubs (e.g. *Angophora*, *Baeckea*, *Corymbia*, *Eucalyptus*, *Leptospermum*, *Melaleuca*), and also flowers of *Bursaria* (Pittosporaceae/Bursariaceae), *Grevillea* (Proteaceae), *Myoporum* (Scrophulariaceae), *Nuytsia floribunda* (Loranthaceae) and *Cassinia* (Asteraceae), but rarely on herbs (e.g. *Temognatha variabilis* on *Actinotus helianthi* [Apiaceae]) (G. Williams pers. obs.).

Known larval host records are from Casuarinaceae and Myrtaceae. From Casuarinaceae: *Temognatha chalcodera* (*Allocasuarina acutivalvis*), *T. fortnumii* (*Allocasuarina vericillata*), *T. heros* (*Casuarina* sp.), *T. martinii* (*Allocasuarina corniculata*), *T. similis* (*Casuarina* sp.), and *T. suturalis* (*Allocasuarina littoralis*, *A. torulosa*, *A. verticillata*). From Myrtaceae: *T. excisicollis* (*Corymbia polycarpa*, *Eucalyptus propinqua*), *T. flavocincta* (*Eucalyptus leucoxylon*), *T. fusca* (*Eucalyptus foecunda*, *E. gracilis*, *E. oleosa*, *E. uncinata*), *T. goryi* (*Eucalyptus amplifolia*, *E. punctata*, *E. tereticornis*), *T. grandis* (*Eucalyptus foecunda*, *E. gracilis*, *E. oleosa*, *E. uncinata*), *Eucalyptus foecunda*, *E. gracilis*, *E. oleosa*, *E. uncinata*, *Melaleuca uncinata*), *T. imperialis* (*Eucalyptus striaticalyx*), *T. marginalis* (*Eucalyptus* spp.), *T. mnizechii* (*Eucalyptus foecunda*, *E. gracilis*, *E. oleosa*, *E. uncinata*), and *T. parvicollis* (*Eucalyptus foecunda*, *E. gracilis*, *E. oleosa*, *E. uncinata*) (Bellamy *et al.* 2013).

Peterson (1996) described and figured the pre-ovipositing behaviour of the Western Australian species *Temognatha bruckii*. Several females of *Temognatha bruckii* were observed (and their actions audible from a distance of 1.5 m) 'charcoal-scraping', this involving individual beetles scraping charcoal particles from the burnt section of a tree stump into the extruded ovipositor. Each beetle repeated this behaviour for periods of 20–30 minutes. Macqueen (1948) earlier reported charcoal in the abdomen of *Temognatha fortnumii* and had observed females similarly scraping charcoal from burnt timber. He considered that eggs were coated with charcoal particles prior to being oviposited, though he did not state as to how this was achieved. Peterson (1996) also observed ovipositing behaviour of a second Western Australian species *Temognatha chalcodera*. Females oviposited about 6–7 cm above ground on living smooth-barked *Allocasuarina acutivalvis* (Casuarinaceae) trunks. Single eggs were laid on the outside of the trunk surface and once oviposited each egg was then completely coated with sand (previously collected by scraping sand from the soil surface into the extended ovipositor), thus forming a dome-shaped hemispherical body cemented firmly to the trunk. Scraping sand from the soil surface has been observed for *Temognatha variabilis*, with individuals moving backwards and forwards over the soil surface for periods >10 minutes, each with the ovipositor extended and the tip of the abdomen bent downwards (G. Williams pers. obs.). Ovipositing behaviour observed for *Temognatha variabilis* by one of us (AMS) mirrors that reported for *Temognatha bruckii*, *T. chalcodera* and *T. fortnumii* by Peterson and Macqueen. Females were observed to fly to dwarf *Allocasuarina* (possibly *Allocasuarina stricta*) growing on the summit of Bald Hill in the Blue Mts, west of Sydney. 'Upon landing these descended along main stems. Eggs were laid at the base of each stem, the beetles simultaneously covering the eggs with a liquefied charcoal mixture. Females were also observed scraping charcoal into their extended abdomen from a dead sapling trunk that still had carbon from an old fire' (A. Sundholm pers. obs.). Gardner (1989) suggested the structure of the stigmoderine ovipositor (and what she termed the 'particle sac') was an adaptation allowing collection of coarse particles which might then be used to coat eggs prior to or during oviposition in the soil or directly upon host plants.

General references. Bellamy (2002, 2003, 2005); Bellamy et al. (2013); Burns and Burns (1992); Carnaby (1987); Carter (1916); Cowie (2001); Gardner (1989); Hawkeswood (1978); Lawrence and Lemann (2019); Matthews (1985); Peterson (1989, 1991, 1996, 2015); Williams (2002, 2020a); Williams and Williams (1983).

SUBFAMILY AGRILINAE
Aaaaba Bellamy, 2013
Figs 503, 504.
Tribe. Coraebini Bedel, 1921

General distribution. Australia. **Australian distribution.** Two species, *Aaaaba fossicollis* and *A. nodosa*, both of which occur in New South Wales, Queensland and Victoria. However, their distribution is constrained by the presence of their obligate larval host vines of the genus *Rubus* (Rosaceae), several of which, such as *R. moluccanus* (previously *R. hillii*), can grow prolifically in mesic forest regrowth following disturbance.

Comments. *Aaaaba* is a member of the Corabaeini, a speciose tribe with a great diversity of genera especially to be found in Africa, including Madagascar (Bellamy 2003). The Australian corabaeine fauna is also diverse and in addition to *Aaaaba* is constituted by *Dinocephalia, Diphucrania, Ethonion, Hypocisseis, Meliboeithon, Neospades, Pachycisseis, Paracephala, Stanwatkinsius, Toxoscelus* and *Synechocera*. In earlier literature the genus was referred to *Alcinous*, but this name proved to be preoccupied (by a genus of Pycnogonida, or 'sea spiders) and so was amended to *Aaaba* (as cited in Bellamy 2002, 2003; Bellamy et al. 2013; Lawrence and Lemann 2019), but this name was also shown to be preoccupied by a genus of sponges and so was further amended to *Aaaaba* (Bellamy 2013).

Adults are usually encountered in spring but may be present until late January. *Aaaaba fossicollis* and *A. nodosa* are commonly encountered on the sunlit margins of wet sclerophyll forest and a variety of lowland and higher altitude rainforest formations, and also in dry sclerophyll forest. They occur in association with leaves and canes of their known *Rubus* host vines. *Aaaaba fossicollis* are small (approx. 5 mm) metallic green beetles that feed on the upper leaf epidermis of *Rubus moluccanus* and *R. parvifolius* and will retreat to the leaf undersurface when disturbed. Adults are commonly found at rest at the base of the leaf lamina (Fig. 503) or along the central leaf vein, and move out radially from resting positions to feed and, in so doing, leave characteristic tell-tale blotches on the leaf surface. Feeding tends to be concentrated within the central section of each leaf. Williams (1983) observed on *Rubus moluccanus* at Lansdowne (northern New South Wales) that beetles progressively moved from plant to plant so that by the end of season most had been attacked, though none appeared to have been adversely affected and exhibited healthy growth by mid-autumn. Adult *Aaaaba nodosa* are black–grey, somewhat larger (approx. 7 mm) than *A. fossicollis*, and frequently fly away when disturbed. Adult *Aaaaba nodosa* has been recorded feeding on the leaves of *Rubus moluccanus, R. parvifolius* and *R. moorei*. Observations on their feeding habits indicated that they fed mainly, often exclusively, on the leaf margins leaving irregular excisions indistinguishable from those caused by other insects (e.g. pergid sawfly larvae). *Rubus moluccanus* is also a larval host for *A. nodosa* (Williams 1983). Adults copulate on the leaf surface, the females then depositing eggs on plant canes. Developing larvae feed upon, and pupate within, the pithy interior.

General references. Bellamy (2002, 2003); Bellamy *et al.* (2013); Burns and Burns (1992); Lawrence and Lemann (2019); Williams (1983, 2002, 2020a).

Agrilus Curtis, 1825

Figs **217–222**; 505, 506.

Tribe. Agrilini Laporte, 1835

General distribution. Widespread with an almost cosmopolitan distribution, but absent from New Zealand and a number of other Pacific Islands. **Australian distribution.** Fifty-four species. Genus occurs in all states including Tasmania, with numerous new species being described from the east coast. The distribution of most is restricted to particular regions of the mainland, with that of only a few species (e.g. *Agrilus assimilis, A. australasiae, A. hypoleucus*) being widespread within the continent. However, historically there were few collections from Central and North-western Australia (Curletti 2001).

Comments. The tribe Agrilini is the most speciose higher buprestid taxon (Bellamy 2002) but the Australian *Agrilus* fauna is depauperate relative to that of other regions, for example the Afrotropical region (600 spp.), Oriental (500 spp.), Palearctic (300 spp.) and New World (900 spp.) (Curletti 2001, 2010). Species are mainly bronze–black or coppery, often with a distinct pubescence or pubescent markings (*Agrilus bispinosus, A. queenslandicus*), but with several notable exceptions that are metallic green or primarily so (e.g. *Agrilus archaicus, A. terraereginae, A. walesicus*). Adults are elongate in form (especially elongate and very slender in *A. deyrollei*) and range in size from approximately 3 mm (*Agrilus bilby, A. numbat*) to over a centimetre in length (*A. bispinosus, A. mastersii*). They inhabit a diversity of habitats including heath and woodland, sclerophyll forest, rainforest, and shrubby dune complexes, *Agrilus carterellus* being an example occurring commonly in the latter habitat. Although several species occupy diverse woodland and open forest communities (e.g. *Agrilus assimilis, A. danesi, A. hypoleucus*) a number are to be found in rainforest (e.g. *A. deauratus, A. decupratus, A. frenchi, A. proserpinae,*

A. walesicus) including littoral rainforest and floristically depauperate and structurally simple vine scrubs.

Adults are herbivorous and are commonly encountered on the foliage of Casuarinaceae and especially *Acacia*. Rarely are individuals found associated with flowers. Adult host plant records include *Agrilus assimilis* (*Acacia ligulata, A. pycnantha* [Fabaceae], *Eucalyptus fasciculosa* [Myrtaceae]); *Agrilus aurovittatus* (*Allocasuarina huegeliana, A. torulosa* [Casuarinaceae]); *Agrilus australasiae* (*Acacia dealbata, A. decurrens, A. parramattensis, A. pycnantha, A. sophorae* [Fabaceae], *Allocasuarina littoralis, Casuarina suberosa* [Casuarinaceae]); *Agrilus decupratus* (*Acacia binervata* [Fabaceae]); *Agrilus echidna* (*Allocasuarina huegeliana, A. acuminata* [Fabaceae]); *Agrilus funebris* (*Acacia melanoxylon* [Fabaceae], and possibly *Castanopsis acuminatissima, Lithocarpus* sp. [Fagaceae]); *Agrilus hypoleucus* (*Allocasuarina muelleriana, Casuarina cunninghamiana* [Casuarinaceae], *Acacia dealbata, A. decurrens, A. mearnsii, A. parramattensis, A. pycnantha, A. sophorae, A. spectabilis* [Fabaceae], *Leptospermum* sp. [Myrtaceae]); *Agrilus koala* (*A. acuminata* [Fabaceae]); *Agrilus mastersii* (*Acacia harpophylla* [Fabaceae]); *Agrilus occipitalis* (*Citrus aurantiaca, C. maxima, C. sinensis* [Rutaceae]); *Agrilus sundholmi* (*Acacia calamifolia* [Fabaceae]); and *Agrilus wallaby* (*Acacia* sp. [Fabaceae]).

Known larval host records include *Agrilus australasiae* (*Acacia chalkeri, A. pycnantha, A. spectabilis* [Fabaceae]); *Agrilus brevis* (*Allocasuarina huegeliana* [Casuarinaceae]); *Agrilus carterellus* (*Claoxylon australe* [Euphorbiaceae]); *A. deauratus* (*Drypetes deplanchei* [Putranjivaceae/ Euphorbiaceae], *Eucalyptus paniculata* [Myrtaceae], *Acacia longifolia*); *A. decupratus* (*Argyrodendron actinophyllum* [Malvaceae/Sterculiaceae]); *A. echidna* (*Allocasuarina campestris* [Casuarinaceae], *Leptospermum* sp. [Myrtaceae]); *A. frenchi* (*Drypetes deplanchei, Argyrodendron actinophyllum*); *A. hypoleucus* (*Acacia longifolia*); *A. indignus* (*Hibiscus tiliaceus* [Malvaceae]); *A. kangaroo* (*Acacia oswaldii*); *A. koala* (*Acacia* sp.); *A. mastersii* (*Acacia leiocalyx*); *A. occipitalis* (*Citrus aurantiaca, C. maxima, C. sinensis* [Rutaceae]); and *A. thylacinus* (*Casuarina* sp. [Casuarinaceae]).

The exotic *Agrilus hyperici* was introduced from France in 1939 to help in the control of the invasive weed St John's Wort, *Hypericum perforatum* (Clusiaceae) (Bellamy 2002). However, surviving beetle populations now persist at only a few locations (Briese 1991).

General references. Bellamy (2002, 2003); Bellamy *et al.* (2013); Burns and Burns (1992); Carter (1924); Cowie

(2001); Curletti (2001, 2002, 2010); Hawkeswood (1978), Lawrence and Lemann (2019); Matthews (1985); Volkovitsh and Hawkeswood (1990); Obenberger (1959); Williams (1985, 2002, 2020a); Williams and Williams (1983).

Anthaxomorphus Deyrolle, 1865
Figs **223**, **224**.

Tribe. Aphanisticini Jacquelin du Val, 1863

General distribution. Africo-tropical, Australasian and Oriental regions. **Australian distribution.** *Anthaxomorphus queenslandicus* recorded from North Queensland (Claudie River).

Comments. *Anthaxomorphus* shares membership of the tribe Aphanasticini with *Aphanisticus* and *Endelus*; however, this may be an unnatural grouping (Williams and Weir 1992), with Bellamy (2002) making a subtribal distinction (i.e. *Anthaxomorphus* [Anthaxomorphina]; *Aphanisticus*, *Endelus* [Aphanisticina]).

A distinctive feature of *Anthaxomorphus* is the greatly enlarged and robust metafemur, a saltatorial character otherwise found, though not enlarged to the same degree, in *Sambus*. This small genus is represented in Australia by a single bronze–black species, *Anthaxomorphus queenslandicus* (length 3.4 mm), the species elongate–rounded when viewed dorsally, and flattened laterally; the holotype was collected at lights in December 1971 (Williams and Weir 1992). Nothing is known of the species life history, but may possibly be a leaf miner. Two southern African species have been recorded from *Ficus* spp. (Moraceae) (Bellamy 1987a).

Elsewhere in the Australian region *Anthaxomorphus* is represented by *A. bougainvillensis* (Bougainville Island), *A. granulosus* and *A. papuanus*; the latter two species collected by Alfred Russel Wallace in western Irian Jaya (Williams and Weir 1992). The type specimen of *Anthaxomorphus bougainvillensis*, figured in Lawrence and Lemann (2019) and Williams and Weir (1992), is distinctly more broadly elongate–ovate than *A. queenslandicus*.

General references. Bellamy (1987a, 2002, 2003); Hołyński (1993); Lawrence and Lemann (2019); Williams and Weir (1992).

Aphanisticus Latreille, 1829
Fig. **225**.

Tribe. Aphanisticini Jacquelin du Val, 1863

General distribution. The genus occurs in the Afro-tropical, Madagascan, Indo-Oriental, Palaearctic and Australasian regions. **Australian distribution.** Three species (*Aphanisticus blackburni*, *A. browni*, *A. endeloides*) are primarily recorded from northern Australia.

Comments. In addition to records from northern Australia *Aphanisticus* is also known from littoral and lowland subtropical rainforest, and estuarine rush–saltmarsh communities in New South Wales. Adults are narrow and minute in size (<3 mm), black or nearly so, and in New South Wales have been collected in Malaise intercept traps in September–January. In form *Aphanisticus* resembles *Endelus* (also a member of the Aphanasticini), and although species of *Endelus* tend to be somewhat broader in general plan, the two genera can be separated principally on the formation of the antennae (e.g. terminal antennal segments forming a club in *Aphanisticus*) (Lawrence and Lemann 2019). There are currently no published host plant records for Australian species, though cylindrically stemmed species of estuary-associated *Juncus* (Juncaceae) and *Isolepis* (Cyperaceae) (Fig. 535), both genera widespread in Australia (Hnatiuk 1990), are potential candidates. European species are recorded utilising reeds and rushes as larval host plants (Bílý 1999), and several species are pests of sugar cane and turfgrass (Chang and Otto 1987; Kang *et al.* 2016).

Selected collection localities: *Aphanisticus blackburni* (South Johnstone River [Qld], *A. endeloides* (North Cairns [Qld]).

General references. Bellamy (2002, 2003); Bílý (1999); Carter (1924); Lawrence and Lemann (2019).

Dinocephalia Obenberger, 1923
Figs **226–228**.

Tribe. Coraebini.

General distribution. Australia. **Australian distribution.** Seven species: *Dinocephalia browni* (WA), *D. burnsi* (Vic), *D. carteri* (SA), *D. cyaneipennis* (Qld, NSW, Vic, Tas), *D. leucogaster* (WA), *D. thoracica* (Qld, NSW, Vic, Tas, SA, WA) and *D. transsecta* (Qld, SA, Vic).

Comments. These are all elongate, somewhat cylindrical beetles ranging in size from approximately 4 (*Dinocephalia browni*) to 12 mm (*D. thoracica*). Species are similar in appearance to *Meliboeithon* and *Paracephala* but can be separated from both these by the loss of the epipleural carinae; additionally from *Paracephala* which has antennae serrate from segment 5 (from segment 4 in *Dinocephalia*) and the absence of a longitudinal groove on the head which is present in *Meliboeithon* (Bellamy 1988). Species exhibit sexual dimorphism: females tend to be larger than males, with additional characters including the pygidium which is narrowed and produced (especially so in *Dinocephalia thoracica*), and the antennae are longer than wide. Species are generally black–bronze with *Dinocephalia cyaneipennis* possessing metallic dark blue elytra. Habitats include tall heath, shrubland and woodland as well as open sclerophyll forest. Adults are commonly associated with Casuarinaceae, with which the genus may have co-evolved, and are often seen in sunlight on foliage. However, individuals often quickly fall to the ground when disturbed. Different species may be found co-habiting on the same plant host, as with *Dinocephalia thoracica* and *D. cyaneipennis* in the Sydney region.

Host records include adult *Dinocephalia cyaneipennis* on *Allocasuarina distyla*, *A. littoralis*, *A. torulosa*, *Casuarina cunninghamiana* (Casuarinaceae), *Dillwynia* spp. and other bush peas (Fabaceae), *Leptospermum polygalifolium* (Myrtaceae) and larvae in galls in *Allocasuarina distyla*; adult *D. burnsi* on foliage of *Casuarina* spp.; and adult *D. thoracica* on foliage of *Allocasuarina distyla* and on flowers of *Bursaria* sp. (Pittosporaceae/Bursariaceae).

General references. Bellamy (1988, 2002, 2003); Bellamy *et al.* (2013); Burns and Burns (1992); Cowie (2001); Hawkeswood (1978); Lawrence and Lemann (2019); Turner and Hawkeswood (1994); Williams (2002); Williams and Williams (1983).

Diphucrania Dejean, 1833
Figs **229–236**; 507–509.
Tribe. Coraebini Bedel, 1921

General distribution. Occurs widely in the Philippines, Indonesia, New Guinea, Australia and some adjacent islands (Barker 2001, 2006b). **Australian distribution.** Comprises 112 species (Lawrence and Lemann 2019), though it is likely more await discovery. The genus is widespread in all states including Tasmania (5 spp.), being especially well recorded from the south-eastern and south-western regions of the mainland.

Comments. Species assigned to *Diphucrania* were previously placed in *Cisseis*; however, the former name has priority. Species range in size from <4 mm (*Diphucrania aenea*, *D. patricia*) to approximately 14 mm (*D. leucosticta*, *D. regale*) and are broadly elongate in form, with the majority being bronze–black. Several have distinct white spots on their elytra (e.g. *Diphucrania duodecemmaculata*, *D. leucosticta*, *D. robertifisheri*, *D. stigmata*), or the pronotum is a metallic bronze or bronze–green that contrasts with the otherwise dull elytra (*Diphucrania leucosticta*, *D.*

regale, *D. tyleri*). There are several instances of species being brightly metallic coloured (e.g. *Diphucrania aeruginosa*, *D. chlorata*). Individual species are to be found in a wide variety of plant communities; however, the genus is more characteristic of xeric habitats such as shrub, woodland and open forest complexes. Few species are associated with rainforest: *Diphucrania williamsi* from northern New South Wales and southern Queensland (Barker 2006b), and possibly *D. watkinsi* (also from northern New South Wales) being apparent rainforest specialists.

Individuals can often be found on foliage, frequently that of *Acacia* (e.g. *Diphucrania acuducta*, *D. aenea*, *D. aureocyanea*, *D. chalcophora*, *D. cupripennis*, *D. fulgidicollis*, *D. tasmanica*) and Casuarinaceae (*D. duodecemmaculata*, *D. rubricatas*, *D. scabiosa*), sometimes both (*Diphucrania cupreicollis*, *D. scabrosula*), but also can be commonly associated with flowers (*D. minutissima*, *D. nitidiventris*, *D. notulata*, *D. oblonga*, *D. obscura*, *D. viridiceps*, *D. vicina*). Although many feed on nectar (e.g. that of *Leptospermum* [Myrtaceae]) there are instances of feeding on actual floral parts (e.g. *D. viridiceps* on *Patersonia occidentalis* [Iridaceae]).

An overview of adult host plants from which various *Diphucrania* species have been recorded (on either flowers or leaves) includes *Allocasuarina campestris*, *A. distyla*, *A. verticillata*, *Casuarina cunninghamiana*, *C. glauca* (Casuarinaceae), *Callitris* sp. (Cupressaceae), *Epacris* sp. (Epacridaceae), *Leucopogon* sp. (Ericaceae), *Acacia aulacocarpa*, *A. cincinnata*, *A. dealbata*, *A. decurrens*, *A. falcata*, *A. filicifolia*, *A. harpophylla*, *A. holosericea*, *A. leiocalyx*, *A. leptostachya*, *A. linifolia*, *A. longifolia*, *A. mangium*, *A. melanoxylon*, *A. obtusifolia*, *A. parramattensis*, *A. pendula*, *A. pycnantha*, *Daviesia latifolia*, *Dillwynia floribunda*, *D. retorta*, *Goodenia* sp., *Jacksonia scoparia*, *Oxylobium* sp., *Phyllota grandiflora*, *P. phylicoides*, *Pultenaea* spp., *Tephrosia astragaloides* and *Viminaria juncea* (Fabaceae), *Glischrocaryon roei* (Haloragaceae), *Patersonia occidentalis* (Iridaceae), *Angophora hispida*, *Baeckea imbricata*, *Chamelaucium uncinatum*, *Eucalyptus* spp., *Leptospermum coriaceum*, *L. polygalifolium*, *L. myrtifolium*, *L. trinervium*, *Kunzea ericoides*, *Melaleuca* sp., *Syzygium floribundum*, *S. smithii*, *Tristaniopsis* sp. (Myrtaceae); *Bursaria* sp. (Pittosporaceae/Bursariaceae), *Banksia attenuata*, *Grevillia* sp., *Hakea teretifolia* (Proteaceae), *Dodonaea triquetra* (Sapindaceae), *Xanthorrhoea johnsonii*, *X. preissii* (Xanthorrhoeaceae) and *Xyris* sp. (Xyridaceae) (Bellamy *et al.* 2013).

Larval host records are relatively few: *Diphucrania acuducta* (*Bossiaea rhombifolia*, *Dillwynia retorta* and galls in roots of *Pultenaea stipularis* [Fabaceae]); *D. cupripennis* (*Acacia longifolia* [Fabaceae]); *Diphucrania cyanea* (*Acacia blakelyi*); *Diphucrania duodecimmaculata* (*Xanthorrhoea* spp. [Xanthorrhoeaceae]); and *D. williamsi* (*Elattostachys nervosa* [Sapindaceae]) (Bellamy *et al.* 2013).

General references. Barker (1999, 2001, 2002, 2006b); Bellamy (2002, 2003); Bellamy *et al.* (2013); Burns and Burns (1992); Carter (1923a); Cowie (2001); Hawkeswood (1978); Kubán *et al.* (2001); Lawrence and Lemann (2019); Matthews (1985); Williams (2002, 2020a); Williams and Williams (1983).

Endelus Deyrolle, 1865
Fig. **237**.

Tribe. Aphanisticini Jacquelin du Val, 1863

General distribution. Distributed in Oriental and Australasian regions, including Fiji and various other islands in the Western Pacific. **Australian distribution.** Represented only by *Endelus subcornutus*, which occurs in Queensland and New Guinea (Bellamy 2002).

Comments. *Endelus* shares with the related genus *Aphanisticus* the deep excavation between the eyes but is more broadly elongate in form to the latter. Extralimital species are small in size, generally about 4 mm in length, and black or metallic-hued. Elsewhere species are reported as associated with, and bred from, ferns. However, there are no known Australian hosts.

General references. Bellamy (2002, 2003); Lawrence and Lemann (2019); Théry (1932).

Ethonion Kubán, 2001
Figs **238–242**.

Tribe. Coraebini Bedel, 1921

General distribution. Australia. **Australian distribution.** Eight species, the genus widely distributed in mainland Australia and Tasmania: *Ethonion breve* and *E. roei* (WA), *E. corpulentum* (NSW, SA, Vic, Tas), *E. fissiceps* (NSW, Vic), *E. jessicae* (NSW), *E. leai* (SA, Vic, Tas), *E. maculatum* (Qld, NSW, SA, Vic) and *E. reichei* (Qld, NSW, Vic, Tas).

Comments. Species previously placed in *Ethon*, and in overall form resemble certain *Diphucrania*, though generally more broadly bodied and with the head possessing a narrow, elongate longitudinal groove on the frontovertex. A key to species is given in Hawkeswood and Turner (1994). Individual species range in size from approximately 5 (*Ethonion fissiceps*, *E. leai*) to about 11 mm (*E. corpulentum*) in length, but there is variation in size within species. Adults are often black–bronze with a whitish pubescence scattered on the elytra (*Ethonion corpulentum*, *E. leai*, *E. maculatum*, *E. reichei*). In contrast *Ethonion fissiceps* is metallic bronze. Species inhabit heath, shrubland, and open woodland and sclerophyll forest habitats, and are frequently found associated with flowering Fabaceae.

Adult host plant records include *Ethonion breve* on flowers of *Patersonia occidentalis* (Iridaceae); *Ethonion corpulentum* on flowers of *Daviesia latifolia*, *Dillwynia* spp., *Pultenaea acerosa*, *P. daphnoides*, *Pultenaea* spp. (Fabaceae) and *Leptospermum* spp. (Myrtaceae); *E. fissiceps* on flowers and foliage of *Dillwynia floribunda*; *Ethonion leai* on flowers and foliage of *Dillwynia retorta* and *D. sericea*; *E. maculatum* on flowers of *Jacksonia* sp. (Fabaceae); *Ethonion* sp. on flowers of *Xyris* sp. (Xyridaceae), and *E. reichei* on flowers of *Dillwynia floribunda*, *D. retorta*, *Dillwynia* spp., *Jacksonia scoparia*, *P. ferruginea*, *Pultenaea* spp. and *Leptospermum* spp.. With the exception of *Leptospermum* adult plant hosts possess yellow (as with *Xyris*) or yellow and red flowers (Fabaceae).

Ethonion species produce galls in the stems and branches, and sometimes roots, of Fabaceae. Stem and/or branch gall host records include *Ethonion corpulentum* (in *Dillwynia retorta*), *E. fissiceps* (*Dillwynia glaberrima*, *D. retorta*, *Pultenaea blakelyi*, *P. flexilis*), *E. jessicae* (*Dillwynia glaberrima*, *D. retorta*, *Pultenaea blakelyi*, *P. flexilis*), and *E. reichei* (*Pultenaea ferruginea*, *P. flexilis*, *P. stipularis*). *Ethonion maculatum* has been found in root galls of *Dillwynia retorta*.

General references. Bellamy (2002, 2003); Bellamy *et al.* (2013); Burns and Burns (1992); Carter (1923a); Cowie (2001); Hawkeswood and Turner (1994); Lawrence and Lemann (2019); Kubán *et al.* (2001); Matthews (1985); Williams (2002); Williams and Williams (1983).

Germarica Blackburn, 1887
Figs **243**; 510.

Tribe. Aphanisticini Jacquelin du Val, 1863

General distribution. Australia. **Australian distribution.** There are three described species: *Germarica blackburni* (Qld), *G. carteri* (Qld) and *G. lilliputana* (Qld, NSW, SA, Vic, Tas). However, there are potentially undescribed species (Bellamy 2002; G. Williams unpubl. data).

Comments. Previously placed in Trachyinae (Cobos 1979). In general appearance species of *Germarica* closely resemble *Cylindromorphus* (Palaearctic), *Paracylindromorphus* (Africa, Madagascar, Oriental and Palaearctic regions) and *Cylindromorphoides* (Neotropics) (Bellamy 2003). Species historically assigned to *Germarica lilliputana* represent a complex of species (G. Williams unpubl. data). Two species described by H. J. Carter (1926) as *Germarica* (*G. abbreviata* [WA], *G. elata* [Qld]) have been transferred to *Helferella* (Bellamy and Peterson 2000a). Beetles are minute (≤3 mm in length), cylindrical and slender in form, and lustrous black. Adults are found on the foliage of *Casuarina* and *Allocasuarina* species (Fig. 510), the genus possibly having a co-evolutionary association with Casuarinaceae.

Germarica lilliputana occurs in woodland, dry sclerophyll forest and estuarine-associated mixed *Casuarina*-rush swamps. Bellamy *et al.* (2013) list adult *Germarica lilliputana* on foliage of *Allocasuarina distyla*, *A. littoralis*, *A. nana*, *A. verticillata* and *Casuarina suberosa* (Casuarinaceae). The species has also been recorded from *Allocasuarina torulosa* and *Casuarina glauca* (G. Williams unpubl. records). Larval hosts are unknown but may be associated with the needle-like leaves of Casuarinaceae (Lawrence and Lemann 2019).

General references. Bellamy (2002, 2003); Bellamy *et al.* (2013); Burns and Burns (1992); Carter (1926); Cobos (1979); Cowie (2001); Hawkeswood (1978); Lawrence and Lemann (2019); Matthews (1985); Obenberger (1923b); Williams and Williams (1983).

Habroloma Thomson, 1864

Figs **244**; 511, 512.

Tribe. Trachyini Laporte, 1835

General distribution. Africa, Madagascar, Oriental, Australasian and Palaearctic regions. **Australian distribution.** Seven species are recorded from Australia: *Habroloma australasiae* (Qld), *H. australis* (WA), *H. clythia* (Qld), *H. crataeis* (Qld), *H. frenchi* (Qld, WA), *H. pauperula* (NSW) and *H. socialis* (NSW). However, the fauna may be greater and the genus is in need of revision.

Comments. Small to minute species (approx. 1.8–3 mm), elongate–ovate to almost triangular in form, bronze to bronze–black, sometimes with distinct whitish pubescence on the elytra. Similar in form to *Trachys* (a single species [*T. blackburni*] of this genus recorded from Australia [Bellamy 2002]) but these two genera can be best separated by each elytron in *Habroloma* possessing a longitudinal carina that extends from the humeral callus towards the apex. The elytral carinae are absent in *Trachys*.

Adults have been collected between September and April, and are recorded from woodland, open sclerophyll forest and tropical and subtropical rainforest. An undetermined species has been collected from riparian rainforest in the Northern Territory (Douglas River, south-west of Hayes Creek) (G. Williams unpubl. data).

Adult *Habroloma australis* have been found on leaves of *Cantharospermum reticulatum* (Fabaceae), and those of *H. socialis* are commonly found feeding on the upper epidermis of *Kennedia rubicunda* (Fig. 512) and often on those of *Desmondium* spp. leaves (Fabaceae). Tell-tale pale blotches on the upper leaf surfaces of these two genera characterise adult feeding activity. Adults when disturbed will quickly fall from the leaves upon which they are feeding, but will frequently return to the same leaf. The larvae of *Habroloma socialis* are also leaf miners of *Kennedia rubicunda*. In addition adult *Habroloma socialis* have been commonly observed on the leaves of *Elaecarpus obovatus* (Elaeocarpaceae) and occasionally *Podocarpus elatus* (Podocarpaceae) in subtropical littoral rainforest in New South Wales (G. Williams unpubl. data). Lea (1895) writes of *Habroloma socialis* (as *Trachys socialis*): 'I obtained numerous specimens on a vine (the leaves of which were perfectly riddled by them) in thick scrub [rainforest]) on the banks of the Little River (a tributary of the Clarence [NSW]'. An undetermined *Habroloma* species has been collected from *Macaranga* (Euphorbiaceae) in northern Queensland (J. Hasenpusch pers. comm.). The exotic species *Habroloma vansoni* is recorded from the leaves of *Combretum* (Combretaceae) in South Africa (C. Bellamy unpubl. records). Bílý *et al.* (2008) record a myrmecophilous association between the larvae of the exotic species *Habroloma myrmecophila* and ants of the genus *Oecophylla*: however, myrmecophily in Australian *Habroloma* species has not been reported.

General references. Bellamy (2002, 2003); Bellamy *et al.* (2013); Bílý *et al.* (2008); Carter (1928a); Cobos (1979); Hawkeswood (2007b); Williams (2002, 2020a).

Hedwigiella Obenberger, 1941

Fig. **245**.

Tribe. Trachyini Laporte, 1835.

General distribution. Neotropics. **Australian distribution.** One introduced species (*Hedwigiella jureceki*) (Qld).

Comments. *Hedwigiella jureceki* superficially resembles *Paratrachys* and particularly *Pachyschelus*; however, *Paratrachys* are smaller (no larger than 3.5 mm) and *Pachyschelus* differs in details of antennae and general body structure, as well as distribution (north-west WA) (Lawrence and Lemann 2019). Previously in the genus *Hylaeogena* and a native of South America, *Hedwigiella jureceki* was introduced into Australia to help control the rampant invasive vine *Macfadyena unguis-cati* (Bignoniaceae), a species commonly known as 'cat's claw creeper' (Snow and Dhileepan 2014), and now naturalised in disturbed rainforest north from southern New South Wales (Harden *et al.* 2007).

General references. Bellamy (2003); Hespenheide (2014); Lawrence and Lemann (2019).

Hypocisseis Thomson, 1879
Figs **246**, **247**; 513, 514.
Tribe. Coraebini Bedel, 1921.
General distribution. Indo-Australasian region from Indonesia and the Philippines to Australia. **Australian distribution.** Ten species: *Hypocisseis brachyformis*, *H. cyanura*, *H. latipennis*, *H. madari* (Qld), *H. pilosicollis* (Qld, NSW), *H. ornata* (Vic) and *H. suturalis* (Qld, NSW, Vic, WA), with the distribution of three species described by Obenberger (1924a) (i.e. *H. blackburni*, *H. carteri*, *H. nigrosericea*) being uncertain.

Comments. Adults in general form resemble *Diphucrania* and *Ethonion*. A number of species were previously placed, sometimes diversely, in *Cisseoides* (e.g. *Hypocisseis blackburni*, *H. carteri*, *H. cyanura*, *H. madari*, *H. nigrosericea*), *Ethon* (*H. latipennis*), *Cisseis* (*H. brachyformis*, *H. suturalis*), *Coraebus* (*H. pilosicollis*) and *Maschalix* (*H. latipennis*). Adults are elongate–ovate in form and range in size from about 6 mm (*Hypocisseis suturalis*) to about 12 mm (*H. latipennis*), are blackish with some species (*Hypocisseis pilosicollis*, *H. suturalis*) with distinct areas of white colouration (formed by densely packed setae) on the pronotum. The genus is recorded from eucalypt-dominated wet sclerophyll forest and dry sclerophyll forest. Adult beetles can be encountered from spring to early autumn, but most commonly from November to January.

Adults of *Hypocisseis latipennis* have been recorded on foliage of *Alphitonia excelsa* (Rhamnaceae), those of *H. ornata* on *Amyema* spp. (Loranthaceae), *H. pilosicollis* on leaves of *Maytenus silvestris* (Celastraceae), and *H. suturalis* on foliage of *Casuarina* spp. (Casuarinaceae),

Jacksonia sp. (Fabaceae), *Leptospermum polygalifolium* (Myrtaceae), *Hakea* spp. (Proteaceae) and *Exocarpos cupressiformis* (Santalaceae). Larvae of *Hypocisseis suturalis* have also been recorded from *Exocarpos cupressiformis*.

General references. Bellamy (1988, 2002); Bellamy *et al.* (2013); Burns and Burns (1992); Carter (1923a, 1928a); Lawrence and Lemann (2019); Obenberger (1924a).

Meliboeithon Obenberger, 1920
Figs **248**; 515.
Tribe. Coraebini Bedel, 1921.
General distribution. Australia. **Australian distribution.** Six species: *Meliboeithon bicostatum* (Qld, NSW, SA), *M. confusum* (WA), *N. crassum* (Qld), *M. cylindricolle* (NSW, SA, WA), *M. intermedium* (Qld, NSW, Vic) and *M. vitticeps* (WA).

Comments. Species are variously bronze, black, or black with purple or golden reflections. Adults are elongate–ovoid and parallel-sided, that in general cylindrical form resemble *Paracephala* and *Dinocephalia*, but have a distinguishing narrow, elongate longitudinal groove on the frontovertex of the head (as also in *Ethonion*), in addition to other characters given in Bellamy 1988. They range in approximate size from <5 to 9 mm, though individuals within species (e.g. *Meliboeithon confusum*, *M. intermedium*) can vary considerably in size, and inhabit open woodland, sclerophyll forest, estuarine wetlands and semi-arid environments. Species are variously encountered from about November to early autumn, with most records being from November and especially December.

Adults are generally collected by sweeping foliage, or even in pitfall traps (e.g. *Meliboeithon cylindricolle*). Bellamy (1988) opined that the collection of *Meliboeithon cylindrinicolle* in ground traps (e.g. Fowlers Gap [NSW], Banjiwarn, Yundamindra [WA]), as was the case with individuals of the type series, suggested '... a larval biology which utilises the root crown or stem base of the host plant, with emergent adults flying very low to feeding and mating sites'.

Meliboeithon intermedium, the most widely distributed species (occurring in hinterland and coastal localities), has been commonly collected in New South Wales during December to March, by sweeping *Juncus* (Juncaceae) and *Isolepis* (Cyperaceae) (Fig. 515) plants growing in tidally inundated coastal wetlands (Fig. 535).

Adults feed on the stems; however, these are too narrow for *Juncus* or *Isolepis* to serve as larval hosts. An additional undetermined *Meliboeithon* sp. has been collected from swellings on the lower part of the stems of *Oxylobium aciculiferum* (Fabaceae) (Hawkeswood 2006b).

General references. Bellamy (1988, 2002, 2003), Bellamy *et al.* (2013); Hawkeswood (2006b); Lawrence and Lemann (2019).

Neospades Blackburn, 1887
Figs **249**, **250**; 516.
Tribe. Coraebini Bedel, 1921

General distribution. Australia. **Australian distribution.** Eleven described species, occurring in all mainland states and territories but especially well represented in Queensland (9 spp.). Apparently absent from Tasmania.

Comments. *Neospades* are small elongate and parallel-sided beetles, approximately 3.5–6 mm in length. Adults resemble *Diphucrania* in general appearance, are dull or metallic-hued, though sometimes bright in colour like the green and black *Neospades chrysopygius*, a species recorded from Queensland, New South Wales, Victoria and Western Australia. Species are active in late spring and early summer, and have been recorded from heath–shrub complexes, woodland, wet sclerophyll forest, lowland subtropical rainforest and araucarian vine scrub.

Adult host records include *Neospades chrysopygius* on flowers and foliage of *Cantharospermum reticulatum*, *Jacksonia thesioides* (Fabaceae), *Sida cordifolia* (Malvaceae) and *Solanum lasiophyllum* (Solanaceae), *N. cruciatus* on flowers of *Hibiscus rhodopetalus* (Malvaceae), *N. lateralis* on leaves of *Acacia aulacocarpa* (Fabaceae), *N. simplex* on flowers of *Corymbia polycarpa* and *Eucalyptus drepanophylla* (Myrtaceae) and *Jacksonia thesioides*; and *N. viridis* on foliage of *Tephrosia astragaloides* (Fabaceae). An enigmatic *Neospades*-like species (Fig. 250) from northern New South Wales has been reared (emerging in November, December) from the narrow branch tips of the small rainforest tree *Diospyros australis* (Ebenaceae), the tips being killed as a result of larval activity (G. Williams unpubl. data).

General references. Bellamy (2002, 2003); Bellamy *et al.* (2013); Lawrence and Lemann (2019).

Pachycisseis Théry, 1929
Fig. **251**.
Tribe. Coraebini Bedel, 1921

General distribution. Australia. **Australian distribution.** A single species, *Pachycisseis bicolor*, recorded from the Northern Territory, Queensland, New South Wales, Victoria and South Australia.

Comments. *Pachycisseis bicolor* was previously placed in *Ethon* (see Bellamy 2002) and keyed with *Cisseis* by Carter (1923a). It occurs in habitats ranging from coastal heath-shrubland complexes to open eucalypt forest of the montane hinterland. Beetles are approximately 8 mm in length, broadly ovate–elongate, with a bronze pronotum and lustrous blue–black elytra collectively marked with about 6–8 irregular white dots formed by very short setae. There is a moderately deep concavity between the eyes but the frontvertex is not longitudinally furrowed as in the similar appearing genus *Ethonion*. In New South Wales adults are active in December, January and February and have been recorded on leaves and flowers of *Acacia spilleriana*, leaves of *A. holosericea* (Fabaceae), leaves and flowers of *Leptospermum polygalifolium* (Myrtaceae), and the flowers of a 'grey-flowered' *Grevillea* sp. (Proteaceae) (G. Williams pers. obs.).

General references. Bellamy (2002, 2003); Bellamy *et al.* (2013); Burns and Burns (1992); Lawrence and Lemann (2019); Williams and Williams (1983).

Pachyschelus Solier, 1833
Fig. **252**.
Tribe. Trachyini Laporte, 1835

General distribution. Nearctic, Neotropical and Oriental regions. **Australian distribution.** An undescribed, or described Oriental, species is recorded from Broome, Western Australia (Plate 81b in Lawrence and Lemann 2019).

Comments. The Western Australian specimen was collected in November. Species of *Pachyschelus* are known to be leaf miners (Queiroz (2002). Adult *Pachyschelus* resemble *Paratrachys*, *Hedwigiella* and some *Habroloma* in form, they are small to minute in size (2–3 mm), blackish (sometimes glossy), and broadly ovate in shape. *Croton* spp. (Euphorbiaceae), some Fabaceae, and species of Cecropiaceae are larval hosts for exotic *Pachyschelus* (Hespenheide and Bellamy 2004; Lawrence and Lemann 2019).

General references. Bellamy (2003); Lawrence and Lemann (2019).

Paracephala Saunders, 1868
Figs **253**; 517.
Tribe. Coraebini Bedel, 1921

General distribution. Australia. **Australian distribution.** Seven species: *Paracephala aenea* (NT, WA), *P. borea* (NT, WA), *P. deserta* (WA), *P. hesperia* (WA), *P. murina* (Qld, NSW, Vic), *P. orientalis* (NT, WA) and *P. pistacina* (Qld, NSW, Vic, SA).

Comments. Adults are slender, parallel-sided and subcylindrical in shape, small (3.5–9 mm) in length, and in general elongate form resemble *Dinocephalia* and *Meliboeithon*. Individual species are black (*Paracephala occidentalis*), black with metallic brassy green reflections (e.g. *P. pistacina*, *P. deserta*, *P. hesperia*, *P. borea*), or black with coppery reflections (*P. aenea*, *P. murina*).

Species are associated with grasslands and have been recorded from early spring to autumn. Adults occur on grasses: *Paracephala borea* and *P. deserta* on flowers of *Stipa* (Poaceae), *P. pistacina* on *Stipa setacea* stems, and *P. murina* on flowers of *Chrysopogon fallax* (Poaceae) (Bellamy 1988; Bellamy *et al.* 2013). *Paracephala murina* has also been recorded on the foliage of *Allocasuarina distyla* (Casuarinaceae).

General references. Bellamy (1988, 2002, 2003); Bellamy *et al.* (2013); Burns and Burns (1992); Goudie (1920); Lawrence and Lemann (2019); Matthews (1985).

Sambus Deyrolle, 1865
Figs 254, 255
Tribe. Coraebini Bedel, 1921.

General distribution. Afro-tropical, Oriental and Australasia regions, including several islands in the Pacific but not New Zealand. **Australian distribution.** One species, *Sambus australis* (South Johnstone River, NQld).

Comments. The single known Australian species *Sambus australis* was previously described as *Cisseis* (*Hypocisseis*) *minuta* by Carter (1923a) (Bellamy and Peterson 2000a). There are about 150 species of *Sambus* worldwide. Species are small, parallel-sided, frequently black with metallic hues, and elytra with zig-zag patterned pale pubescence. The hind femur is swollen (Fig. 255), a saltatorial character also found in *Anthaxomorphus* (Fig. 224). Adults of the Philippine species *Sambus isis* are associated with *Ficus* (Moraceae) (Jackman 1987); Australian hosts are unknown.

General references. Bellamy (2002, 2003); Bellamy and Peterson (2000a); Carter (1923a); Lawrence and Lemann (2019).

Stanwatkinsius Barker & Bellamy, 2001
Figs 256; 518, 519.
Tribe. Coraebini Bedel, 1921.

General distribution. Australia. **Australian distribution.** Sixteen species widely distributed in mainland Australia, especially so in Western Australia, but apparently absent from Tasmania (Bellamy 2002).

Comments. Species somewhat resemble *Diphucrania*, but females possess an ovipositor closer in structure to *Meliboeithon* (Bellamy 2002), this being formed from incurved setae (see discussion in Barker and Bellamy 2001). As in *Diphucrania* the adults are mainly foliage feeders and the larvae bore in decaying timber (Barker and Bellamy 2001). Adults are elongate and rounded towards the apex, are small and range in size from a little less than 6 mm (*Stanwatkinsius cinctus*, *S. demarzi*, *S. uniformis*) to over 8 mm (*S. careniceps*), but as with many other buprestid species females tend to be larger than males. No species have spots on the elytra (as in a number of *Diphucrania*). Adults are generally metallic coppery purple, green, blue or bronze-hued, sometimes with contrasting marginal colours (*Stanwatkinsius grevilleae*, *S. lindi*) or prominently bicolorous, as in *S. crassus*. In some species individuals exhibit a variety of colour forms (e.g. as with *Stanwatkinsius perplexus*), and in others (*S. careniceps*) males and females have characteristically different colours. Individual populations may exhibit distinctive colour varieties (*S. powelli*). Species are especially active during September–November and occupy woodland and open forest. Individuals are usually collected by beating or sweeping foliage, but there are instances of beetles being collected in intercept traps; an example is *Stanwatkinsius demarzi*, a species apparently particular to *Banksia menziesii* woodland in Western Australia (Barker and Bellamy 2001).

In addition to the association of *Stanwatkinsius demarzi* with *Banksia menziesii* mentioned above adult host plant records include *Stanwatkinsius amanda* on foliage of *Grevillea juncifolia* (Proteaceae), *S. careniceps* on foliage of *Allocasuarina campestris* (Casuarinaceae), *S. cinctus* on foliage of *Hakea* spp. (Proteaceae), *S. constrictus* on foliage of *Allocasuarina humilis*, *Allocasuarina* sp., *Grevillea* sp. and *Hakea* spp., *S. crassus* on flowers of *Grevillea* sp., *S. grevilleae* on foliage of *Allocasuarina campestris*, *Grevillea* sp. and *Hakea* sp., *S. kermeti* on foliage of *Casuarina cunninghamiana* (Casuarinaceae), *S. perplexus* on foliage of *Allocasuarina helmsii* and *Allocasuarina* sp.,

S. subcarinifrons on foliage of *Allocasuarina* spp., and *S. uniformis* on foliage of *Allocasuarina verticillata*. In addition, in western New South Wales (Round Hill Nature Reserve) *Stanwatkinsius lindi* has always been found on flowering *Hakea leucoptera* but in the Big Desert region of north-western Victoria adults are associated with *Hakea mitchelli* (A. Sundholm pers. obs.). Collectively, these records suggest a co-evolutionary relationship with Proteaceae and Casuarinaceae, this differing from the association of many *Diphucrania* with *Acacia* (Fabaceae) and other plant taxa.

General references. Barker and Bellamy (2001); Bellamy (2002, 2003); Lawrence and Lemann (2019).

Synechocera Deyrolle, 1865

Figs **257**; 520.

Tribe. Coraebini Bedel, 1921.

General distribution. Australia, but with an enigmatic record for *Synechocera deplana* from the island of Amboina (Bellamy 1987b, 2002). **Australian distribution.** Ten species (?), the genus occurring in all states including a single species (*Synechocera deplana*) from Tasmania (Bellamy 2002). *Synechocera deplana* is widespread in south-eastern Australia and ranges from coastal lowlands to montane localities above 1000 m. The distribution of the remaining species is more localised (see Bellamy 1987b).

Comments. Species are small (4–<9.5 mm), elongate and parallel-sided, markedly dorsoventrally flattened, a morphological adaptation to a compressed lifestyle (Bellamy 1987b). There is a single deep longitudinal groove on the frontovertex, and adults possess a notably rounded pronotum which is widest at about the middle. *Synechocera* inhabits woodland and open forest and montane eucalypt forest, as well as lowland shrub complexes and heathland where *Gahnia* (Cyperaceae) larval food plants occur. Most species are dull black, except for *Synechocera cyanipennis* (velvety blue elytra contrasting with a black head, pronotum and undersurface) and *S. bicolor* (elytra black with coppery reflections). *Aulacostethus rubidus* wasps (Aulacidae) are recorded as parasites of *Synechocera deplana* (as *S. tasmanica* in Bellamy 1987b) larvae. The flattened *Synechocera* body plan allows adults to retreat to the base of the densely packed leaf stems of their *Gahnia* and *Xanthorrhoea* host plants, making their observation and capture difficult (see Fig. 520).

Adult host plant records include *Synechocera brooksi* and *S. queenslandica* on *Xanthorrhoea minor* (Xanthorrhoeaceae), *S. parvipennis* on *Xanthorrhoea* sp., and *S. deplana* on foliage of *Gahnia sieberiana*, *Acacia longifolia* (Fabaceae) and *Leptospermum* sp. (Myrtaceae). However, the *Acacia* and *Leptospermum* records may be fortuitous findings of itinerant individuals. *Synechocera deplana* larvae have been recorded from the flower stalks of *Gahnia* spp. (Cyperaceae). Larvae mine the woody part of the stalk and pack the frass into the central stalk (which is naturally hollow), the bright yellow frass produced indicating larval activity. Several larvae may be present in each stalk. Larvae generally mine upwards towards the seed head but may retrace their tunnelling in a downward direction; this suggests females lay their eggs near the base of the stalks. Adults emerge within 12 months from flower stalks that have set seed the previous summer (Bellamy 1987b). Bellamy also suggested that larvae of *Synechocera* might be found in the stalks of *Xanthorrhoea*.

General references. Bellamy (1987b, 2002, 2003); Bellamy and Peterson (2000a); Bellamy *et al.* (2013); Carter (1924, 1928a); Cowie (2001); Kubán *et al.* (2001); Lawrence and Lemann (2019); Matthews (1985); Théry (1923a).

Toxoscelus Deyrolle, 1865

Fig. **258**.

Tribe. Coraebini Bedel, 1921.

General distribution. North-eastern Palaearctic, Oriental and Australasian regions. **Australian distribution.** Two species: *Toxoscelus eylandtus* (Groote Eylandt, NT) and *T. queenslandicus* (vic. of Mt Webb, NQld).

Comments. Individuals are small in size (approx. 5 mm), elongate and parallel-sided, dull black with metallic reflections, and with zig-zag patterned, pale pubescence on the elytra. In general the appearance of adult *Toxoscelus* species resembles Australian *Diphucrania* and *Sambus*, as well as a number of foreign coraebine genera (e.g. *Anadora*, *Angatra*, *Coraebus*). The holotype female *Toxoscelus queenslandicus* was collected in early October; the host and habitat preferences of both Australian species are unknown (Bellamy and Peterson 2000b).

General references. Bellamy (2002, 2003); Bellamy and Peterson (2000b).

Trachys Fabricius, 1801

Fig. **259**.

Tribe. Trachyini Laporte, 1835.

General distribution. *Trachys* is widely recorded from the Afro-tropical, Oriental, Palaearctic and Australasian regions (Bellamy 2003). **Australian distribution.** A single enigmatic species, *Trachys blackburni*, is recorded from Australia (New South Wales).

Comments. Numerous species earlier assigned to *Trachys* (Carter 1928a) have been moved to *Habroloma* (Bellamy 2002). The remaining Australian taxon, *Trachys blackburni*, resembles *Habroloma* in appearance but the two genera can be separated by the absence of a longitudinal carina on each elytron (present in *Habroloma*) as well as additional characters cited in Lawrence and Lemann's (2019) key to the genera of the Australian buprestid fauna; though the species is 'sub-triangular' in form, not clearly triangular, or 'more or less oval' as their key couplet suggests. Kerremans (1896), in his description of *Trachys blackburni*, simply cites the collection locality as 'New South Wales'. Lawrence and Lemann (2019) also note *Trachys blackburni* as occurring in Papua New Guinea. In Australia the species is possibly associated with littoral rainforest but host and habitat preferences are unknown. European *Trachys* are recorded as leaf miners of a wide range of dictyledonous plant genera and families (Lawrence and Lemann 2019).

General references. Bellamy (2002, 2003), Bílý (1999); Carter (1928a); Kerremans 1896; Lawrence and Lemann (2019).

COLOUR SECTION 2: LIVE BEETLES*

Fig. 265. *Astraeus* sp. cf. *aberrans*. ©Allen M. Sundholm

Fig. 268. *Astraeus hanloni*. ©Allen M. Sundholm

Fig. 266. *Astraeus dedariensis*. ©Allen M. Sundholm

Fig. 269. *Astraeus lineatus*. ©Allen M. Sundholm

Fig. 267. *Astraeus dilutipes*. ©Geoff Williams

Fig. 270. *Astraeus major*. ©Allen M. Sundholm

---

*   Size variation within genera is discussed for each genus in Chapter 2.

Fig. 271. *Astraeus mastersii.* ©Allen M. Sundholm

Fig. 274. *Astraeus sundholmi.* ©Allen M. Sundholm

Fig. 272. *Astraeus multinotatus.* ©Allen M. Sundholm

Fig. 275. *Astraeus williamsi.* ©Allen M. Sundholm

Fig. 273. *Astraeus pygmaeus.* ©Allen M. Sundholm

Fig. 276. *Astraeus williamsi* (lateral). ©Allen M. Sundholm

Fig. 277. *Astraeus yarrattensis.* ©Geoff Williams

Fig. 278. *Astraeus yarrattensis.* ©Geoff Williams

Fig. 279. *Prospheres aurantiopictus.* ©Geoff Williams

Fig. 280. *Xyroscelis bumanna.* ©Geoff Williams

Fig. 281. *Xyroscelis crocata.* ©Allen M. Sundholm

Fig. 282. *Chalcophorotaenia australasiae.* ©Allen M. Sundholm

Fig. 283. *Chalcophorotaenia beltanae.* ©Allen M. Sundholm

Fig. 286. *Chalcophorotaenia cuprascens.* ©Allen M. Sundholm

Fig. 284. *Chalcophorotaenia castanea.* ©Allen M. Sundholm

Fig. 287. *Chalcophorotaenia elongata.* ©Allen M. Sundholm

Fig. 285. *Chalcophorotaenia cerata.* ©Allen M. Sundholm

Fig. 288. *Chalcophorotaenia laeta.* ©Allen M. Sundholm

Fig. 289. *Chalcophorotaenia violacea.* ©Allen M. Sundholm

Fig. 292. *Pseudotaenia gigas.* ©Allen M. Sundholm

Fig. 290. *Cyphogastra pistor.* ©Allen M. Sundholm

Fig. 293. *Pseudotaenia salamandra.* ©Allen M. Sundholm

Fig. 291. *Paracupta lambertii.* ©Geoff Williams

Fig. 294. *Pseudotaenia spilota.* ©Allen M. Sundholm

Fig. 295. *Pseudotaenia waterhousei.* ©Allen M. Sundholm

Fig. 298. *Bubastes carteri.* ©Allen M. Sundholm

Fig. 296. *Anilara* sp. cf. *mephisto.* ©Allen M. Sundholm

Fig. 299. *Bubastes erbani.* ©Allen M. Sundholm

Fig. 297. *Anilara* sp. ©Allen M. Sundholm

Fig. 300. *Bubastes magnifica.* ©Allen M. Sundholm

Fig. 301. *Bubastes subnigricollis* (m). ©Allen M. Sundholm

Fig. 304. *Euryspilus* sp. ©Allen M. Sundholm

Fig. 302. *Cyrioides australis* (mating) on *Banksia spinulosa* (Proteaceae). ©Geoff Williams

Fig. 305. *Melobasis aurocyanea.* ©Allen M. Sundholm

Fig. 303. *Diadoxus regius.* ©Allen M. Sundholm

Fig. 306. *Melobasis gloriosa cruentata* (m). ©Allen M. Sundholm

Fig. 307. *Melobasis gratiosissima* (m). ©Allen M. Sundholm

Fig. 310. *Melobasis purpurilata*. ©Allen M. Sundholm

Fig. 308. *Melobasis janae*. ©Allen M. Sundholm

Fig. 311. *Melobasis pyritosa*. ©Allen M. Sundholm

Fig. 309. *Melobasis nobilitata*. ©Allen M. Sundholm

Fig. 312. *Nascioides costatus*. ©Geoff Williams

Fig. 313. *Nascioides falsomultesimus.* ©Geoff Williams

Fig. 316. *Nascioides tillyardi.* ©Geoff Williams

Fig. 314. *Nascioides pulcher.* ©Geoff Williams

Fig. 317. *Neobubastes aureocincta.* ©Allen M. Sundholm

Fig. 315. *Nascioides quadrinotatus* (lateral). ©Simon Grove (TMAG)

Fig. 318. *Neobubastes nickerli.* ©Allen M. Sundholm

Fig. 319. *Neobuprestis williamsi.* ©Geoff Williams

Fig. 322. *Selagis olivacea* (m). ©Allen M. Sundholm

Fig. 320. *Selagis aurifera.* ©Geoff Williams

Fig. 323. *Selagis viridicyanea* (f). ©Allen M. Sundholm

Fig. 321. *Selagis intercribrata.* ©Allen M. Sundholm

Fig. 324. *Torresita cuprifera.* ©Geoff Williams

Fig. 325. *Calodema plebeium.* ©Allen M. Sundholm

Fig. 328. *Castiarina abdita.* ©Allen M. Sundholm

Fig. 326. *Calodema plebeium* (lateral). ©Allen M. Sundholm

Fig. 329. *Castiarina abdominalis.* ©Allen M. Sundholm

Fig. 327. *Calodema regale* on *Tristaniopsis laurina* (Myrtaceae). ©Geoff Williams

Fig. 330. *Castiarina acuminata.* ©Geoff Williams

Fig. 331. *Castiarina aeraticollis.* ©Allen M. Sundholm

Fig. 334. *Castiarina armstrongi* (variation). ©Allen M. Sundholm

Fig. 332. *Castiarina alternecosta.* ©Geoff Williams

Fig. 335. *Castiarina armstrongi* (variation). ©Allen M. Sundholm

Fig. 333. *Castiarina andersoni.* ©Allen M. Sundholm

Fig. 336. *Castiarina attenuata.* ©Allen M. Sundholm

Colour section 2: Live beetles   111

Fig. 337. *Castiarina aurantiaca.* ©Allen M. Sundholm

Fig. 340. *Castiarina bifasciata.* ©Geoff Williams

Fig. 338. *Castiarina aureola.* ©Allen M. Sundholm

Fig. 341. *Castiarina binotata.* ©Allen M. Sundholm

Fig. 339. *Castiarina beatrix.* ©Geoff Williams

Fig. 342. *Castiarina brooksi.* ©Allen M. Sundholm

Fig. 343. *Castiarina brutella* (lateral). ©Geoff Williams

Fig. 346. *Castiarina carinata.* ©Allen M. Sundholm

Fig. 344. *Castiarina bugejiana.* ©Allen M. Sundholm

Fig. 347. *Castiarina carnabyi.* ©Allen M. Sundholm

Fig. 345. *Castiarina burnsi.* ©Allen M. Sundholm

Fig. 348. *Castiarina castelnaudi.* ©Allen M. Sundholm

Fig. 349. *Castiarina cincta.* ©Allen M. Sundholm

Fig. 352. *Castiarina costipennis.* ©Allen M. Sundholm

Fig. 350. *Castiarina cinnamomea.* ©Allen M. Sundholm

Fig. 353. *Castiarina crocicolor.* ©Allen M. Sundholm

Fig. 351. *Castiarina clancula.* ©Allen M. Sundholm

Fig. 354. *Castiarina cruenta.* ©Allen M. Sundholm

Fig. 355. *Castiarina crux.* ©Allen M. Sundholm

Fig. 358. *Castiarina daedalea.* ©Allen M. Sundholm

Fig. 356. *Castiarina cupreoflava.* ©Allen M. Sundholm

Fig. 359. *Castiarina daranj.* ©Allen M. Sundholm

Fig. 357. *Castiarina cyanipes.* ©Allen M. Sundholm

Fig. 360. *Castiarina decipiens.* ©Allen M. Sundholm

Fig. 361. *Castiarina delta.* ©Geoff Williams

Fig. 364. *Castiarina discoidea.* ©Allen M. Sundholm

Fig. 362. *Castiarina desideria.* ©Allen M. Sundholm

Fig. 365. *Castiarina distincta.* ©Allen M. Sundholm

Fig. 363. *Castiarina dimidiata.* ©Allen M. Sundholm

Fig. 366. *Castiarina distinguenda.* ©Allen M. Sundholm

Fig. 367. *Castiarina eborica.* ©Geoff Williams

Fig. 370. *Castiarina erythroptera.* ©Geoff Williams

Fig. 368. *Castiarina erubescens.* ©Allen M. Sundholm

Fig. 371. *Castiarina euclae.* ©Allen M. Sundholm

Fig. 369. *Castiarina erythromelas.* ©Allen M. Sundholm

Fig. 372. *Castiarina* sp. cf. *flava* (variation). ©Allen M. Sundholm

Fig. 373. *Castiarina* sp. cf. *flava* (variation). ©Allen M. Sundholm

Fig. 376. *Castiarina garrawillae.* ©Allen M. Sundholm

Fig. 374. *Castiarina flavopicta.* ©Simon Grove (TMAG)

Fig. 377. *Castiarina imitator.* ©Geoff Williams

Fig. 375. *Castiarina frauciana.* ©Allen M. Sundholm

Fig. 378. *Castiarina gentilis.* ©Geoff Williams

Fig. 379. *Castiarina gordonburnsi.* ©Allen M. Sundholm

Fig. 382. *Castiarina hanloni.* ©Allen M. Sundholm

Fig. 380. *Castiarina goudiana.* ©Allen M. Sundholm

Fig. 383. *Castiarina harrisoni.* ©Geoff Williams

Fig. 381. *Castiarina gracilior.* ©Allen M. Sundholm

Fig. 384. *Castiarina humeralis.* ©Geoff Williams

Fig. 385. *Castiarina hypocrita.* ©Allen M. Sundholm

Fig. 388. *Castiarina impressicollis.* ©Allen M. Sundholm

Fig. 386. *Castiarina ignea.* ©Allen M. Sundholm

Fig. 389. *Castiarina indigesta.* ©Allen M. Sundholm

Fig. 387. *Castiarina ignota.* ©Allen M. Sundholm

Fig. 390. *Castiarina indigoventricosa.* ©Allen M. Sundholm

Fig. 391. *Castiarina insculpta* ovipositing on *Ozothamnus hookeri*. ©Simon Grove (TMAG)

Fig. 394. *Castiarina insularis*. ©Simon Grove (TMAG)

Fig. 392. *Castiarina insculpta*. ©Simon Grove (TMAG)

Fig. 395. *Castiarina intacta*. ©Allen M. Sundholm

Fig. 393. *Castiarina insignis*. ©Geoff Williams

Fig. 396. *Castiarina interstitialis*. ©Allen M. Sundholm

Fig. 397. *Castiarina jubata* ©Simon Grove (TMAG)

Fig. 400. *Castiarina kirbyi.* ©Allen M. Sundholm

Fig. 398. *Castiarina jubata* (lateral). ©Simon Grove (TMAG)

Fig. 401. *Castiarina laevinotata.* ©Allen M. Sundholm

Fig. 399. *Castiarina kerremansi.* ©Allen M. Sundholm

Fig. 402. *Castiarina lepida.* ©Allen M. Sundholm

Fig. 403. *Castiarina livida.* ©Allen M. Sundholm

Fig. 406. *Castiarina luteocincta.* ©Allen M. Sundholm

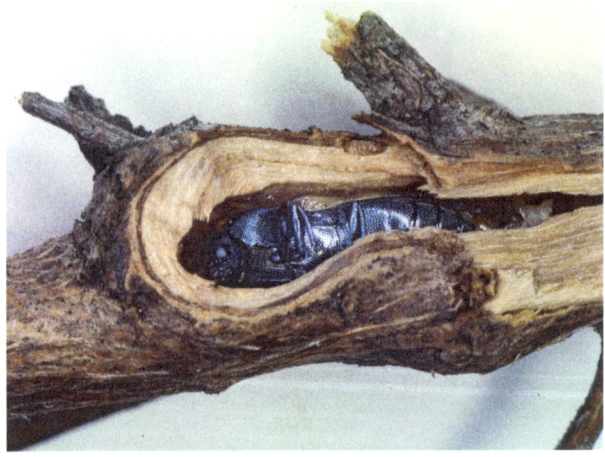

Fig. 404. *Castiarina livida,* pharate adult in *Ozothamnus diosmifolius* stem. ©Allen M. Sundholm

Fig. 407. *Castiarina luteofusca.* ©Allen M. Sundholm

Fig. 405. *Castiarina luteipennis.* ©Geoff Williams

Fig. 408. *Castiarina lycida.* ©Allen M. Sundholm

Fig. 409. *Castiarina macarthuri.* ©Allen M. Sundholm

Fig. 412. *Castiarina maculipennis.* ©Geoff Williams

Fig. 410. *Castiarina macquillani.* ©Simon Grove (TMAG)

Fig. 413. *Castiarina magnificollis.* ©Allen M. Sundholm

Fig. 411. *Castiarina maculicollis.* ©Allen M. Sundholm

Fig. 414. *Castiarina marginicollis.* ©Allen M. Sundholm

Fig. 415. *Castiarina mimesis.* ©Allen M. Sundholm

Fig. 418. *Castiarina neglecta* on *Avicennia marina* (Acanthaceae). ©Geoff Williams

Fig. 416. *Castiarina minuta.* ©Geoff Williams

Fig. 419. *Castiarina obliqua.* ©Geoff Williams

Fig. 417. *Castiarina mustelamajor.* ©Allen M. Sundholm

Fig. 420. *Castiarina obsepta.* ©Allen M. Sundholm

Fig. 421. *Castiarina octomaculata.* ©Allen M. Sundholm

Fig. 424. *Castiarina pallidipennis.* ©Allen M. Sundholm

Fig. 422. *Castiarina femorata.* ©Geoff Williams

Fig. 425. *Castiarina pearsoni.* ©Allen M. Sundholm

Fig. 423. *Castiarina ornata.* ©Allen M. Sundholm

Fig. 426. *Castiarina picta.* ©Allen M. Sundholm

Fig. 427. *Castiarina* sp. cf. *piliventris*. ©Allen M. Sundholm

Fig. 430. *Castiarina producta*. ©Geoff Williams

Fig. 428. *Castiarina placens*. ©Allen M. Sundholm

Fig. 431. *Castiarina prolata*. ©Allen M. Sundholm

Fig. 429. *Castiarina praetermissa*. ©Geoff Williams

Fig. 432. *Castiarina puella*. ©Allen M. Sundholm

Fig. 433. *Castiarina pulchripes*. ©Geoff Williams

Fig. 436. *Castiarina rubicunda*. ©Allen M. Sundholm

Fig. 434. *Castiarina recta* (melanic form). ©Allen M. Sundholm

Fig. 437. *Castiarina rudis*. ©Simon Grove (TMAG)

Fig. 435. *Castiarina robusta*. ©Allen M. Sundholm

Fig. 438. *Castiarina rufipennis*. ©Allen M. Sundholm

Fig. 439. *Castiarina rufipennis,* pharate adult in *Acacia terminalis.* ©Allen M. Sundholm

Fig. 442. *Castiarina sanguinolenta.* ©Allen M. Sundholm

Fig. 440. *Castiarina rufolimbata.* ©Allen M. Sundholm

Fig. 443. *Castiarina scalaris.* ©Geoff Williams

Fig. 441. *Castiarina rutila.* ©Allen M. Sundholm

Fig. 444. *Castiarina semicincta.* ©Allen M. Sundholm

Fig. 445. *Castiarina serratipennis.* ©Allen M. Sundholm

Fig. 448. *Castiarina sieboldi.* ©Allen M. Sundholm

Fig. 446. *Castiarina sexguttata.* ©Allen M. Sundholm

Fig. 449. *Castiarina simulata* (eastern form). ©Allen M. Sundholm

Fig. 447. *Castiarina sexplagiata.* ©Allen M. Sundholm

Fig. 450. *Castiarina skusei.* ©Allen M. Sundholm

Fig. 451. *Castiarina spinolae.* ©Geoff Williams

Fig. 454. *Castiarina suttoni.* ©Allen M. Sundholm

Fig. 452. *Castiarina stellata.* ©Allen M. Sundholm

Fig. 455. *Castiarina tasmaniensis.* ©Simon Grove (TMAG)

Fig. 453. *Castiarina* sp. cf. *sundholmi.* ©Allen M. Sundholm

Fig. 456. *Castiarina tigris.* ©Allen M. Sundholm

Fig. 457. *Castiarina* sp. cf. *tyrrhena* ©Geoff Williams

Fig. 460. *Castiarina variegata* (m). ©Geoff Williams

Fig. 458. *Castiarina uncata*. ©Allen M. Sundholm

Fig. 461. *Castiarina variegata* (f). ©Geoff Williams

Fig. 459. *Castiarina vanderwoudeae*. ©Allen M. Sundholm

Fig. 462. *Castiarina venusta*. ©Allen M. Sundholm

Fig. 463. *Castiarina virginea.* ©Simon Grove (TMAG)

Fig. 466. *Metaxymorpha grayii* on *Syzygium floribundum* (Myrtaceae). ©Geoff Williams

Fig. 464. *Castiarina viridolinea.* ©Allen M. Sundholm

Fig. 467. *Stigmodera gratiosa* (green-gold form). ©Allen M. Sundholm

Fig. 465. *Castiarina woodi.* ©Allen M. Sundholm

Fig. 468. *Stigmodera jacquinotii.* ©Allen M. Sundholm

Fig. 469. *Castiarina neglecta* attempting to mate with *Stigmodera macularia*. ©Geoff Williams

Fig. 472. *Temognatha carpenteriae* (inland form). ©Allen M. Sundholm

Fig. 470. *Temognatha affinis*. ©Allen M. Sundholm

Fig. 473. *Temognatha fallaciosa*. ©Allen M. Sundholm

Fig. 471. *Temognatha acquilonia*. ©Allen M. Sundholm

Fig. 474. *Temognatha flavicollis* (f). ©Allen M. Sundholm

Fig. 475. *Temognatha flavocincta.* ©Allen M. Sundholm

Fig. 478. *Temognatha gigas.* ©Allen M. Sundholm

Fig. 476. *Temognatha fortnumii* (f). ©Allen M. Sundholm

Fig. 479. *Temognatha lessonii.* ©Allen M. Sundholm

Fig. 477. *Temognatha gemmelli* on *Leptospermum* sp. (Myrtaceae). ©Geoff Williams

Fig. 480. *Temognatha macqueenii.* ©Allen M. Sundholm

Fig. 481. *Temognatha marginalis.* ©Allen M. Sundholm

Fig. 484. *Temognatha mitchellii* (colour variation). ©Allen M. Sundholm

Fig. 482. *Temognatha menalcas.* ©Allen M. Sundholm

Fig. 485. *Temognatha nickerli* (colour variation). ©Allen M. Sundholm

Fig. 483. *Temognatha mitchellii.* ©Simon Grove (TMAG)

Fig. 486. *Temognatha nickerli* (colour variation). ©Allen M. Sundholm

Fig. 487. *Temognatha regia* (variation). ©Allen M. Sundholm

Fig. 490. *Temognatha sanguinea* (colour variation).
©Allen M. Sundholm

Fig. 488. *Temognatha regia* (variation). ©Allen M. Sundholm

Fig. 491. *Temognatha* sp. cf. *sanguineocincta.* ©Geoff Williams

Fig. 489. *Temognatha regia* (variation). ©Allen M. Sundholm

Fig. 492. *Temognatha sanguinipennis.* ©Allen M. Sundholm

Fig. 493. *Temognatha sanguiniventris* (colour variation). ©Allen M. Sundholm

Fig. 496. *Temognatha variabilis* (melanic form) mating on *Actinotus helianthi* (Apiaceae). ©Geoff Williams

Fig. 494. *Temognatha spencii* (mating). ©Geoff Williams

Fig. 497. *Temognatha variabilis*. ©Geoff Williams

Fig. 495. *Temognatha suturalis*. ©Geoff Williams

Fig. 498. *Temognatha varicollis*. ©Allen M. Sundholm

Fig. 499. *Temognatha westwoodii.* ©Allen M. Sundholm

Fig. 502. *Temognatha* sp. ©Allen M. Sundholm

Fig. 500. *Temognatha wimmerae.* ©Allen M. Sundholm

Fig. 503. *Aaaaba fossicollis* on *Rubus moluccanus.* ©Geoff Williams

Fig. 501. *Temognatha* sp. nov. ©Allen M. Sundholm

Fig. 504. *Aaaaba nodosus.* ©Geoff Williams

Fig. 505. *Agrilus* sp. aff. *carterellus.* ©Allen M. Sundholm

Fig. 508. *Diphucrania furfurosa.* ©Geoff Williams

Fig. 506. *Agrilus sundholmi.* ©Allen M. Sundholm

Fig. 509. *Diphucrania leucosticta.* ©Geoff Williams

Fig. 507. *Diphucrania duodecimmaculata.* ©Geoff Williams

Fig. 510. *Germarica* sp. cf. *lilliputana* on foliage of *Allocasuarina littoralis* (Casuarinaceae). ©Geoff Williams

Fig. 511. *Habroloma* sp. ©Geoff Williams

Fig. 514. *Hypocisseis suturalis*. ©Geoff Williams

Fig. 512. *Habroloma* sp. feeding on *Kennedia rubicunda* (Fabaceae). ©Geoff Williams

Fig. 515. *Meliboeithon intermedium* on *Isolepis* (Cyperaceae) stem. ©Geoff Williams

Fig. 513. *Hypocisseis pilosicollis*. ©Geoff Williams

Fig. 516. *Neospades rugiceps*. ©Allen M. Sundholm

Fig. 517. *Paracephala* sp. ©Allen M. Sundholm

Fig. 520. *Synechocera deplana* (lateral). ©Simon Grove (TMAG)

Fig. 518. *Stanwatkinsius* sp. cf. *lindi* (m). ©Allen M. Sundholm

Fig. 521. *Rhinotia* sp. (Belidae). ©Geoff Williams

Fig. 519. *Stanwatkinsius* sp. cf. *lindi* (f). ©Allen M. Sundholm

Fig. 522. *Ellipsidion humerale* (Ectobiidae-Blattodea). ©Geoff Williams

3   Regional buprestid faunas

Since the first decades of the 19th century the Australian jewel beetle fauna has been the subject of fascination and scientific enquiry, yet there has been comparatively little attempt to tabulate the nature of local or regional faunas; Hawkeswood (1978) and Nikitin (1979) are two of the infrequent examples. Even state-wide surveys are few, though those of Burns and Burns (1992), Cowie (2001) and Lang (2020) do allow the conflation of quadrant records. The fauna of the diverse landscapes and vegetation communities of the Northern Territory (e.g. see Figs 561–564) remains poorly investigated.

Despite the popular interest in the group, early collections rarely included label data that informed about the nature of host relationships or the vegetation types in which specimens were collected. The consequence is that published information relevant to a finer understanding of the distribution, composition and conservation of species is not readily available; that which there is resides primarily in institutional databases and museum cabinets. But as Williams and Williams (1983) highlight, the efforts of early collectors in amassing collections of species were not matched by an equal attention to labelling, resulting in a poor understanding of the early buprestid fauna, and representations 'of species within collections [being] often inversely proportional to their natural frequency; the more common species usually being ignored in the field and those less frequently encountered, consistently procured. Thus old museum data may offer scant insight into the faunal diversity and frequency in the area'.

Here is drawn together information on several regional and state-wide faunas, admittedly far from complete, and thus 'snapshot' overviews if you will. It is hoped that these will engage a wider appreciation of Australia's unique buprestid diversity, and will encourage future surveys, in particular for existing reserves and for areas faced with threats of land clearing, urbanisation and fragmentation.

NORTH QUEENSLAND AND THE WET TROPICS

For the purpose of this section the region encapsulates the Cape York Peninsula and areas of rainforest south to the vicinity of Mackay (Figs 523–525). It comprises open savannah woodlands, floristically complex lowland rainforests (e.g. Cape Tribulation), largely continuous highland rainforests of the Wet Tropics World Heritage Area, as well as isolated smaller areas of rainforest to the north (e.g. Iron Range) and south (Paluma). The Wet Tropics World Heritage Area, some 894 400 ha in area, extends for greater than 400 km from the vicinity of Townsville to Cooktown, and experiences an annual average rainfall of 1200–8000 mm; this falls mostly from November to April. The landscape of the World Heritage Area includes some of Australia's highest mountains (e.g. Mount Bellenden Ker, Mount Bartle Frere), its diverse geomorphology resulting in a complex of habitat 'islands' which have been conducive to the evolution of a rich mosaic of endemic plants and animals. These include numerous short-range endemics and a notable element that comprises species with ancient Gondwanan affinities. The buprestid fauna (see Table 1) comprises species associated with open forest and a second group restricted to rainforest. A third cohort is one of species that have been collected in both biomes and that respond to available blossom resources, but in the absence of knowledge of larval host plants, their true habitat affiliation remains uncertain. Not surprisingly, the region's relatively intact rainforest communities (about 80 per cent of that at European occupation remains [Adam 1992]) contain a rich buprestid fauna, with new species still being encountered. The rainforest fauna exhibits high levels of endemicity at the species level; however, there are genera with distinctive Australo-Papuan and Oriental affinities (e.g. *Belionota, Cyphagastra, Calodema, Lampetis, Metaxymorpha, Strigoptera, Toxoschelus*). Members of the large and resplendent stigmoderine genera *Calodema*

and *Metaxymorpha* have long been the attention of entomologists and collectors alike, their size, colour and generally localised distribution making them subjects of commercial, often illicit, trade. Nylander (2008) lists *Calodema regale, C. plebeium, C. rubrimarginata* and *C. wallacei* from Australia. *Calodema regale* (Figs 119–121), a species with bright yellow elytra and conspicuous red 'eye spots' on the pronotum, is restricted to eastern Australia but is the most widely distributed species, its range extending from Cape York Peninsula to the mid-north coast of New South Wales. *Calodema wallacei*, in which the elytra are starkly contrasted yellow and black, has an extralimital distribution in Papua New Guinea and Irian Jaya but in Australia is apparently restricted to isolated sites in the upper Cape York Peninsula, with few specimens known. *Calodema plebeium* (Figs 325, 326) is restricted to the lower eastern region of Cape York Peninsula. Its elytra are red or orange with a distinct wide black fascia behind the middle, this reaching the suture and lateral margins. A

Table 1. Examples (given alphabetically) of buprestid species from Tropical Queensland.

Agrilus aborigines (South Johnstone River), *Agrilus bilby* (Babinda), *Agrilus bispinosus* (Johnstone Riv., Mossman), *Agrilus brevis* (Cape York, Johnstone Riv.), *Agrilus cairnsensis* (Babinda, Kuranda, Mount Spec), *Agriilus deyrollei* (Cairns), *Agrilus fassatii* (Bamaga), *Agrilus indigenus* (Cairns), *Agrilus macleayi* (Cairns), *Agrilus mole* (Davies Creek), *Agrilus moloch* (Kuranda), *Agrilus numbat* (Kuranda)

Anthaxomorphus queenslandicus (Claudie River)

Austrochalcophorella subfasciata (Cape York, Innisfail)

Belionota prasina (Cairns)

Bubastes hasenpuschi (Claudie River, Mt Garnet dist., Newcastle Range)

Calodema plebeium (Cairns, Garradunga, Kuranda), *Calodema regale* (Cairns, Innisfail, Kuranda), *Calodema rubrimarginata* (Julatten, Rex Range), *Calodema wallacei* (Cape York)

Castiarina aeruginosa (Mt Carbine, Mt Molloy), *Castiarina aglaia* (Paluma, Cardwell Range, Kuranda), *Castiarina andersoni* (Paluma, Mareeba), *Castiarina athertonensis* (Mt Carbine, Georgetown, Herberton), *Castiarina atra* (Paluma), *Castiarina aeruginosa* (Mt Molloy), *Castiarina biguttata* (Mt Spec, Mt Carbine), *Castiarina binotata* (Cairns, Herberton), *Castiarina brooksi* (Paluma), *Castiarina campestris* (Cooktown), *Castiarina capensis* (Cape York, Bamaga), *Castiarina carinata* (Cape York, Paluma), *Castiarina chrysothoracica* (Mt Carbine, Georgetown), *Castiarina decipiens* (Cape York), *Castiarina dispar* (Herberton), *Castiarina distincta* (Paluma, Mt Molloy), *Castiarina doddi* (Kuranda, Paluma), *Castiarina erubescens* (Mt Molloy, Cairns), *Castiarina garnettensis* (Mt Molloy, Kuranda), *Castiarina guttifera* (Kuranda), *Castiarina hasenpuschi* (vic. of Innisfail), *Castiarina hawkeswoodi* (Kuranda, Cardwell Range), *Castiarina hemizostera* (Cardwell Range), *Castiarina hoblerae* (Mareeba), *Castiarina hypocrita* (Mt Spec, Mt Carbine, Kuranda), *Castiarina impressicollis* (Cairns), *Castiarina indigesta* (Atherton Tableland), *Castiarina jackhasenpuschi* (Cardwell Range), *Castiarina lauta* (Paluma), *Castiarina luteofusca* (Mt Lewis), *Castiarina magnificollis* (Mareeba), *Castiarina mimica* (Kuranda), *Castiarina nebula* (Kuranda, Mt Lewis), *Castiarina nigriceps* (Mt Molloy), *Castiarina nigriventris* (Magnetic Is.), *Castiarina octosignata* (Kuranda, Lake Eacham), *Castiarina oedemerida* (Georgetown), *Castiarina pallas* (Endeavour River, Thursday Island), *Castiarina paulhasenpuschi* (Croydon), *Castiarina phaeorhaea* (Ayr, Kuranda, Mt Carbine), *Castiarina prolata* (Cardwell Range), *Castiarina puella* (vic. of Innisfail), *Castiarina quadriplagiata* (Normanton), *Castiarina rollei* (Cairns, Windsor Tableland), *Castiarina seminigra* (Cape York), *Castiarina storeyi* (Mt Carbine), *Castiarina suehasenpuschae* (Georgetown), *Castiarina sundholmi* (Paluma), *Castiarina tenebrosa* (Paluma), *Castiarina terraereginae* (Cairns, Mareeba), *Castiarina testudocaput* (Turtle Head Island-Cape York Peninsula), *Castiarina tigris* (Atherton Tableland), *Castiarina titania* (Endeavour River, Kuranda, Mt Molloy), *Castiarina trispiculis* (Paluma, Kuranda, Herberton), *Castiarina tropica* (Cape York), *Castiarina vallisi* (Paluma), *Castiarina venusta* (Cape York, Paluma), *Castiarina viridiventris* (Mt Molloy, Cooktown), *Castiarina walfordi* (Paluma), *Castiarina warningensis* (Cooktown), *Castiarina woodi* (Mt Lewis).

Chalcophorotaenia cuprascens (Kuranda), *Chalcophorotaenia elongata* (Proserpine) *Chrysobothris queenslandica* (Townsville)

Cyphogastra farinosa (Iron Range), *Cyphogastra insolens* (Cairns?), *Cyphogastra pistor* (Cape York Penisula, Townsville), *Cyphogastra vulnerata* (vic. Cairns)

Cyrioides cincta (Kuranda), *Cyrioides sexspilota* (vic. Innisfail)

Dinocephalia thoracica (Cairns, Cape York Peninsula)

Diphucrania brooksi (Mareeba), *Diphucrania carterella* (Kuranda, Herberton), *Diphucrania borealis* (Kuranda, Cape York, vic. Herberton, vic. Mt Carbine), *Diphucrania cupreola* (Mt Spec, vic. Paluma), *Diphucrania excelsior* (vic. Mt Carbine), *Diphucrania pulleni* (vic. Daintree), *Diphucrania stellata* (Paluma, Cairns, vic. Croydon, vic. Lockerbie, Cardwell, Bloomfield River, Cape York), *Diphucrania storeyi* (Mareeba), *Diphucrania suehasenpuschi* (vic. Innisfail), *Diphucrania ustulata* (Cooktown)

Helferella gothmogoides (Tolga), *Helferella macalpinei* (Kuranda), *Helferella tolgae* (Tolga), *Helferella webbensis* (Mt Webb)

Iridotaenia bellicosa (Garradunga, Kuranda)

Maoraxia storeyi (Cape Tribulation, Kuranda)

Melobasis aenea (Kuranda), *Melobasis australis* (NEQld), *Melobasis intricata* (Cape York), *Melobasis kaszabi* (Mt Molloy, Endeavour Riv., Kuranda, Mt Molloy, Prince of Wales Island), *Melobasis macleayi* (vic. Innisfail), *Melobasis quadrinotatus* (Townsville), *Melobasis remotioculata* (Garradunga), *Melobasis smaragdifrons* (Coen), *Melobasis spinosa* (Johnstone River)

Merimna atrata (Palm Cove, Jardine River)

Metataenia sp. (South Johnstone River)

Metaxymorpha gloriosa (Cairns, Garradunga, Kuranda, Mossman), *Metaxymorpha hauseri* (Kuranda, vic. Daintree River)

Nascioides elessarellus (vic. Mareeba), *Nascioides mundus* (Cairns, Cape York), *Nascioides storeyi* (Windsor Tablelands), *Nascioides subcostatus* (Windsor Tablelands), *Nascioides walfordorum* (Cairns, Mt Lewis)

Neospades lateralis (Cooktown)

Paracephala murina (Mt Spec)

Paratrachys queenslandica (vic. Mareeba)

Pseudotaenia frenchi (Georgetown, Croydon)

Sambus australis (South Johnstone River)

Selagis corusca (Mt Molloy)

Strigoptera bimaculata

Stanwatkinsius rhodopus (vic. Croydon)

Synechocera brooksi (vic. Coen, vic. Atherton, Kuranda), *Synechocera cyanipennis* (Cairns dist.), *Synechocera queenslandica* (vic. Paluma, Mt Spec)

Temognatha alternata (Herbert River, Kuranda), *Temognathia aquilonia* (Mt Molloy), *Temognatha carpentariae* (Desailly Range, Windsor Tablelands), *Temognatha franca* (Gordonvale), *Temognatha jansonii* (Yarrabah), *Temognatha nickerli* (vic. Kuranda), *Temognatha regia* (vic. Kuranda), *Temognatha viridescens* (Iron Range)

Toxoscelus queenslandicus (Mt Webb)

second black fascia occurs in the anterior third and though reaching the sutural margins does not continue to the lateral margin. The fourth species, *Calodema rubrimarginata*, has a localised distribution in northern Queensland (Nylander 2008), and was described from specimens caught at the Rex Range in February 1993 (Barker 1993b). The elytra are bright yellow except for a black apical patch and red lateral margins, these becoming wider and darker towards the apex. The genus *Metaxymorpha* includes two species, *M. gloriosa* and *M. hauseri* (Figs 175, 176, 178), that exhibit similar ranges extending north and south of Cairns (Nylander 2008). These respectively have yellow

and reddish brown elytra, both have black apical patches but in *Metaxymorpha gloriosa* the lateral margins are reddish somewhat such that the elytra in colour and pattern are reminiscent of *Calodema rubrimarginata*.

Two of the early entomologists and naturalists that stand out in this region are John George Brooks (1910–75) and Frederick Parkhurst Dodd (1861–1937). Frederick Dodd came to be affectionately known as the 'Butterfly Man of Kuranda', and it is tempting to see parallels in his appreciation of Nature with that of the American philosopher Henry Thoreau. Dodd cooperated with many entomologists of his day, and established a large insect collection, buprestids prominent among them, which formed a popular education exhibit. George Brooks practiced as a dentist and in later years lived at Edge Hill near Cairns. His interests spanned Coleoptera, especially Buprestidae, and their host plants. He corresponded widely, hosted collecting trips (especially to Paluma) by visiting entomologists, and published many notes in the *North Queensland Naturalist* (e.g. Brooks 1948).

CENTRAL QUEENSLAND

This subhumid 'brigalow belt' and associated near-coastal zone is situated between the mesic forests and savannah woodlands of the tropical north, the low latitude subtropical closed and open forests of the southeast, and the semi-desert woodlands and mixed grasslands of far western Queensland. In an arbitrary deliniation, in the north it includes the extensive Carnarvon Gorge National Park (approx. 30 000 ha), Cudmore National Park (20 400 ha) and, closer to the coast, the Blackdown Tableland National Park (32 000 ha), with extensive areas of floristic interest further south (e.g. Mount Walsh and Woowoonga national parks) (Figs 526–530). The region possesses extensive open sclerophyll forests, *Acacia harpophylla*-dominated 'brigalow' woodlands, as well as examples of Araucarian vine scrubs (e.g. Goodnight Scrub) and upland mesic eucalypt forests. Expansive areas of native vegetation, however, have been cleared for agriculture and mining, with surviving vegetation often reduced to isolated remnant patches and remaining under threat.

Despite the loss of vegetation the region retains high biodiversity attributes, including that of the Buprestidae (Table 2). For example, of the more than 20 species of *Melobasis* known from the area about two-thirds are endemic (Levey 2012).

Table 2.   Examples (given alphabetically) of buprestid species from Central Queensland.

Agrilus deauratus (Gayndah), *Agrilus mastersii* (Edungalba, Gayndah), *Agrilus raphelisi* (Mount Walsh, Mourangee), *Agrilus thylacinus* (Edungalba)
Anilara obscura (Gayndah)
Araucariana queenslandica (Ban Ban Range, Imbil)
Astraeus adamsi (Edungalba), *Astraeus blackdownensis* (Blackdown Tableland), *Astraeus mastersii* (Gayndah, Humbolt National Park), *Astraeus mourangeensis* (Edungalba), *Astraeus pygmaeus* (Edungalba), *Astraeus simulator* (Edungalba),
Austrochalcophora subfasciata (Edungalba)
Barakula petersonorum (Barakula)
Bubastes achardi (Milmerran), *Bubastes australasiae* (Edungalba, Duaringa, Carnarvon Range, Milmerran), *Bubastes barkeri* (Edungalba, Mourangee, Roma, Springsure, St George), *Bubastes flavocaerulea* (Cunnamulla), *Bubastes magnifica* (Winton dist.), *Bubastes suturalis* (vic. Cloncurry)
Castiarina acuticollis (Edungalba), *Castiarina adamsi* (Edungalba, Milmerran), *Castiarina affabilis* (Blackdown Tableland), *Castiarina alecgemmelli* (Milmerran), *Castiarina andersoni* (Edungalba), *Castiarina atronotata* (Edungalba), *Castiarina blackdownensis* (Blackdown Tableland), *Castiarina confusa* (Edungalba), *Castiarina decipiens* (Edungalba, Carnarvon Range), *Castiarina dingoensis* (Dingo dist.), *Castiarina dispar* (Edungalba, Dingo), *Castiarina distincta* (Milmerran, Dingo, Edungalba, Duaringa), *Castiarina duaringae* (Duaringa, Edungalba), *Castiarina ernestadamsi* (Edungalba), *Castiarina festiva* (Edungalba, Dingo), *Castiarina flavosignata* (Edungalba, Blackdown Tableland), *Castiarina gibbicollis* (Milmerran, Blackdown Tableland), *Castiarina gilberti* (Blackdown Tableland), *Castiarina hasenpuschi* (Mt Lewis, Cardwell Range), *Castiarina haswelli* (Blackdown Tableland, Paluma), *Castiarina hoblerae* (Gayndah, Milmerran), *Castiarina impressicollis* (Edungalba, Gayndah), *Castiarina inermis* (Edungalba), *Castiarina intacta* (Milmerran, Blackdown Tableland), *Castiarina jucunda* (Milmerran, Blackdown Tableland), *Castiarina latipes* (Milmerran), *Castiarina liliputana* (Gayndah), *Castiarina maculicollis* (Milmerran), *Castiarina melasma* (Milmerran), *Castiarina minuta* (Edungalba), *Castiarina nanula* (*Edungalba*), *Castiarina ochreiventris* (Milmerran, Edungalba), *Castiarina pearsoni* (Blackdown Tableland), *Castiarina planipes* (Edungalba, Milmerran), *Castiarina quadriplagiata* (Edungalba), *Castiarina quinquepunctata* (Dingo, Blackdown Tableland, Edungalba), *Castiarina straminea* (Edungalba, Blackdown Tableland), *Castiarina trispiculis* (Milmerran, Blackdown Tableland), *Castiarina violacea* (Edungalba, Milmerran), *Castiarina viridiventris* (Edungalba).
Chalcophorotaenia cerata (Edungalba)
Chrysobothris mastersii (Gayndah), *Chrysobothris saundersii* (Gayndah), *Chrysobothris subsimilis* (Mourangee), *Chrysobothris viridis* (Gayndah)
Dinocephalia thoracica (Gayndah)
Diphucrania cupreata (Isla Gorge), *Diphucrania cupreola* (Blackdown Tableland), *Diphucrania ernestadamsi* (Edungalba), *Diphucrania impressicollis* (Gayndah), *Diphucrania macqueeni* (Milmerran), *Diphucrania pulleni* (Gayndah, Milmerran, Edungalba), *Diphucrania trimentula* (Milmerran), *Diphucrania watkinsi* (vic. Calliope)

Hypocisseis latipennis (Gayndah)

Meliboeithon crassum (Gayndah), *Meliboeithon intermedium* (Gayndah)

Melobasis apicalis (Gayndah), *Melobasis brevimaculata* (Carnarvon Range, Edungalba, Duaringa), *Melobasis costata* (Gayndah), *Melobasis cupreovittata queenslandicus* (Edungalba), *Melobasis elderi* (Goodnight Scrub), *Melobasis kanguluorum* (Edungalba), *Melobasis obscura* (Gayndah), *Melobasis propinqua* (Blackdown Tableland), *Melobasis pusilla* (Edungalba), *Melobasis quadrinotatus* (Milmerran, Edungalba), *Melobasis sordida* (Duaringa, Edungalba, Milmerran), *Melobasis vertebralis vertebralis* (Duaringa, Edungalba, Carnarvon National Park).

Merimna atrata (Cudmore National Park)

Nascioides viridis (Gayndah)

Neocuris gracilis (Gayndah)

Neospades chrysopygia (Gayndah)

Notobubastes orientalis (vic. Mourangee)

Notographus hieroglyphicus (Gayndah), *Notographus sulcipennis* (Gayndah)

Paracephala borea (Edungalba), *Paracephala murina* (Biggenden, Duaringa, Edungalba)

Polycesta mastersii (Edungalba, Gayndah)

Pseudanilara cupripes (Gayndah), *Pseudanilara purpureicollis* (Gayndah)

Pseudotaenia ajax (Edungalba, Tambo), *Pseudotaenia salamandra* (Dingo, vic. Duaringa)

Selagis adamsi (Edungalba), *Selagis atrocyanea* (Edungalba), *Selagis splendens* (Gayndah, Goodnight Scrub)

Temognatha carpentarie (Springsure), *Temognatha obesissima* (Milmerran)

Over the years the region has been subject to visits by numerous entomologists, among the earliest George Masters who in the 19th century collected for Sir William Macleay. His field collecting resulted in the discovery of many new Buprestidae and in 1872 Macleay published, under the title 'Notes on a collection of insects from Gayndah', descriptions of those species that Masters had discovered, naming several in Masters' honour. These included *Agrilus mastersii*, *Astraeus mastersii*, *Castiarina mastersii* (at that time placed in *Stigmodera*), *Chrysobothris mastersii* and *Polycesta mastersii*. However, several entomologists and naturalists were residents. Prominent among these was Ernest Adams, a pastoralist who lived on the cattle station 'Mourangee' at Edungalba, about 150 km inland from Rockhampton. Although he held an interest in insects in general buprestids became a principal focus, this resulting in the discovery of several new species (e.g. *Astraeus adamsi*, *Castiarina ernestadamsi*, *Selagis adamsi*). Over many years Adams corresponded and interacted widely with other entomologists interested in the family, in 1985 in recognition of his research contributions being awarded an Honorary Field Associate title by the University of Adelaide (Barker 2006a). Species cited in Table 2, in part, bear testimony to his efforts.

CENTRAL AND WESTERN NEW SOUTH WALES

This is a southern extension of the Queensland 'brigalow belt' and includes the Pilliga Scrub (>500 000 ha), which encapsulates the largest contiguous example of semi-arid woodland in the state, and in which are found extensive stands of *Callitris glaucophylla*. Other significant reserves include Yathong Nature Reserve (11 600 ha – mallee, cypress and belah [*Casuarina cristata*] woodlands) and Nombinnie Nature Reserve (70 000 ha – mallee, cypress and belah woodlands), Round Hill Nature Reserve (13 630 ha – extensive mallee communities), Warrumbungles National Park (23 300 ha – mixed eucalypt forests and woodland), Mt Kaputar National Park (3 680 000 ha – mixed semi-arid woodland, subalpine heath and eucalypt forests), and Goonoo Goonoo National Park and Goonoo Goonoo State Conservation Area (63 000 ha combined – eucalypt forests and woodlands) (Figs 544–548). The diversity of vegetation types to be encountered represents a mosaic of habitat and host plant opportunities to which buprestids have variously responded, and although particular reserves have been the subject of extensive collecting no comparative fauna records have been published. There are also examples of vine thickets–dry rainforest (e.g. Nandewar Ranges, Planchonella Nature Reserve) (Curran *et al.* 2008), but these have been poorly investigated. However, since European occupation there has been a dramatic clearing of native vegetation. This continues.

Early collecting in the western regions of New South Wales was undertaken by the government entomologist

Walter Wilson Froggatt. Resident at Callubri Station (which gives its name to *Castiarina callubriensis*), Bogan River, was J. W. T. Armstrong who collected numerous new species, the locality 'Bogan River' in Table 3 indicating

Table 3.   Examples (given alphabetically) of buprestid species from Central and Western New South Wales.

Agrilus danesi (vic. Mount Kaputar National Park), *Agrilus hypoleucus* (vic. Sofala, Pilliga East State Forest)
Anilara angusta (Baradine State Forest, Mount Kaputar National Park, Pilliga East State Forest), *Anilara antiqua* (Mount Kaputar National Park), *Anilara longicollis* (Garrawilla), *Anilara purpurascens* (Baradine State Forest, Mount Kaputar National Park, Pilliga East State Forest), *Anilara sulcicollis* (Mount Kaputar National Park)
Astraeus mastersii (Mount Kaputar National Park), *Astraeus hanloni* (14 km west of Euabalong West, Nombinnie Nature Reserve, Yathong Nature Reserve), *Astraeus pygmaeus* (Kwiambal National Park), *Astraeus sundholmi* (Round Hill Nature Reserve)
Austrophorella quadrisignata (Lightning Ridge, 60 km south-east of Narrabri)
Australorhipis aphanochila (Fowlers Gap)
Bubastes achardi (Cobar), *Bubastes barkeri* (Bogan Riv., vic. Cobar, vic. Euabalong West, Nyngan, Round Hill Nature Reserve), *Bubastes barkeri* (28 km N Cobar), *B. flavocaerulea* (Round Hill Nature Reserve, Yathong Nature Reserve); *Bubastes magnifica* (approx. 100 km NNW Brewarrina)
Burnsiellus trisulcata (Bogan River)
Castiarina abdominalis (Mount Kaputar National Park, Roto, Round Hill Nature Reserve); *Castiarina aenicornis* (Round Hill), *Castiarina alecgemmelli* (Coonabarabran, Pilliga Scrub), *Castiarina allensundholmi* (Round Hill Nature Reserve), *Castiarina armstrongi* (Bogan River), *Castiarina atricollis* (Round Hill Reserve), *Castiarina bugejiana* (Bourke dist.), *Castiarina burnsi* (Lachlan Plains), *Castiarina callubriensis* (Bogan River), *Castiarina clancula* (vic. of Matakana), *Castiarina colligens* (Binnaway, Rocky Glen), *Castiarina crenata* (Mount Kaputar National Park), *Castiarina daranj* (Round Hill, Garrawilla, Parkes), *Castiarina decemmaculata* (Warrumbungles National Park), *Castiarina deyrollei* (Pilliga Scrub), *Castiarina duaringae* (Nyngan), *Castiarina flavosignata* (Pilliga Scrub, Round Hill, Hill End), *Castiarina garrawillae* (Pilliga Scrub, Rocky Glen, 21 km east of Parkes), *Castiarina hilaris* (Warrumbungles National Park), *Castiarina humilis* (Coonabarabran), *Castiarina ignea* (Lachlan Plains), *Castiarina jucunda* (Binnaway, Mount Kaputar National Park, Rocky Glen), *Castiarina laevinotata* (Bogan River, Pilliga Scrub), *Castiarina latipes* (Coonabarabran, Mt Kaputar National Park), *Castiarina maculicollis* (Goonoo Goonoo National Park), *Castiarina marginicollis.* (approx. 14 km north-east of Coonabarabran), *Castiarina markhanloni* (Round Hill), *Castiarina mima* (Mount Kaputar National Park), *Castiarina ochreiventris* (Mount Kaputar National Park, Pilliga Scrub), *Castiarina octomaculata* (Pilliga Scrub), *Castiarina octospilota* (Pilliga Scrub), *Castiarina parallipennis* (Cobar, Pilliga Scrub), *Castiarina punctatosulcata* (Mount Kaputar National Park), *Castiarina scalaris* (Pilliga Scrub, Warrumbungles National Park), *Castiarina subpura* (Pilliga Scrub), *Castiarina turneri* (10 km west of Euabalong West); *Castiarina violacea* (Trangie, Wilcannia)
Chalcophorotaenia cerata (Pilliga Scrub, Round Hill Nature Reserve)
Chrysobothris caelatus (Nombinnie Nature Reserve), *C. subsimilis* (Mt Kaputar National Park)
Diadoxus erythrurus (approx. 46 km E of Coonabarabran)
Dinocephalia cyaneipennis (85 km SW Narrabri)
Diphucrania armstrongi (Bogan River), *Diphucrania duodecimmaculata* (Pilliga Scrub), *Diphucrania marmorata* (Mt Kaputar National Park), *Diphucrania patricia* (46 km E of Coonabarabran), *Diphucrania trimentula* (Pilliga Scrub), *Diphucrania viridiceps* (Coonabarabran, Garrawilla, Mount Kaputar National Park)
Meliboeithon cylindricolle (vic. Fowlers Gap)
Melobasis burnsi (vic. Euabalong West), *Melobasis cupreovittata cupreovittata* (Kinchega National Park), *Melobasis paucipunctata* (vic. Euabalong West), *Melobasis propinqua* (Mount Kaputar National Park, Pilliga State Forest, Warrumbungles National Park), *Melobasis sordida* (Gunnedah, Mt Hope), *Melobasis vertebralis cuneata* (vic. Coonabarabran, Condobolin, Cobar, Mt Kaputar National Park, Round Hill Nature Reserve, Warrumbungles National Park)
Nascioides olliffi (Mt Kaputar National Park)
Neocuris cupreolatera (vic. Pilliga Scrub), *Neocuris gracilis* (95 km south-west Narrabri), *Neocuris guerinii* (Pilliga East State Forest)
Paracephala pistacina (Bogan River)
Pseudotaenia ajax (Turrawan), *P. salamandra* (approx. 100 km NNW Brewarrina), *P. waterhousei* (Mt Kaputar National Park, Round Hill, Turrawan)
Selagis caloptera (Mt Kaputar National Park)
Temognatha duponti (Coonabarabran, Dubbo, Pilliga Scrub), *Temognatha fallaciosa* (Goonoo Goonoo National Park), *Temognatha variabilis* (Pilliga East State Forest)

buprestids collected by him. The entomologist H. J. Carter visited the Coonabarabran region in January 1923 in the company of T. G. Sloane, during which he discovered the previously unknown *Castiarina latipes* on flowering *Leptospermum*, and later *Temognatha barbiventris*[1] (see *T. duponti*, Fig. 192) (first collected at Beechworth, Victoria) at the time only the second known specimen of the species, and which he failed to find during further ventures to the region (Carter 1933). He makes note of the capricious seasonal nature of insect occurrence during his visits to the Pilliga Scrub (Fig. 544) and surrounds, the abundance of his favourite beetles always subject to dry weather conditions – and of the long hours of field investigation among flowering plants, yet often with little to show for his efforts.

SYDNEY AND THE BLUE MOUNTAINS

Not surprisingly the Sydney region, comprising the site of the first European settlement, has been subject to a level of collecting likely unsurpassed by any other area in Australia. To date some 200 or so species of Buprestidae have been recorded from the region (e.g. Williams 1977; Hawkeswood 1978; Williams and Williams 1983). Sydney and its environs sit within the much more extensive Sydney Basin, this reaching approximately to Lithgow in the west, Newcastle in the north, and Batemans Bay in the south. This wedge-shaped region comprises some 3 600 000 ha, rising from sea level to altitudes exceeding 1100 m in the western Blue Mountains. Although the dominant rock types weather to poor soils, a rich flora of some 2000 plants has been recorded (Beadle *et al.* 1972). The diversity of buprestids collected reflects this richness of plants.

The Stigmoderinae are well represented, with more than 100 species recorded (in Williams 1977; Hawkeswood 1978; Nikitin 1979; Williams and Williams 1983; Williams 1987), the great majority being of *Castiarina* (>90 species), with a number (e.g. *C. coeruleipes, C. eborica, C. flavopurpurea, C. kanangara, C. kershawi, C. kirbyi, C. klugii, C. leai, C. nasuta, C. skusei, C. variopicta, C. watkinsi*) being more characteristic of elevated sites of the western, or 'Upper', Blue Mountains. Among the species recorded from the Upper Blue Mountains is the enigmatic *Castiarina lisaejessicae*. *Castiarina lisaejessicae* was described by Hawkeswood and Turner (2009) on the basis

of a single male specimen (unfortunately later destroyed) collected from flowering *Leptospermum morrisonii* (Myrtaceae). The species somewhat resembles *Castiarina klugii*, its pronotum being bright metallic blue with lighter blue to blue–green reflections, and the elytra being dark purple with maroon reflections and with two orange fasciae (one at about the middle and a second subapical one). Of the large-sized *Temognatha* recorded, *T. affinis, T. goryi, T. grandis, T. limbata, T. mitchellii, T. sexmaculata, T. thoracica* and *T. vitticollis* are particularly of interest, all species often associated with *Eucalyptus* that flower in mid-late summer; individuals typically observed on the highest of blossoms. In addition to these, Nikitin (1979) notes the rarely reported *Temognatha goryi* (Fig. 194) found on a eucalypt trunk at Cabramatta and a dead specimen excavated from a eucalypt trunk at Fairfield Park, both locations once western suburbs of the sprawling Sydney metropolis. Other Buprestidae of note include *Xyroscelis bumanna, Melobasis placida, Nascio xanthura, N. vetusta, Nascioides carissimus, N. costatus, N. macalpinei, N. parryi, N. pulcher, Cyrioides imperialis, Iridotaenia albivittis* (the last on the trunks of young *Eucalyptus* trees) and *Castiarina armata*, the last considered threatened in the Sydney region by clearing and development (Turner and Hawkeswood 1996a).

If the Fabaceae–Mimosoideae are considered separately from the Fabaceae–Faboideae, four major plant groups (adding the Casuarinaceae and Myrtaceae) characterise the majority of plant–buprestid associations in the Sydney Basin. Though these hold dominance in most of Australia's xeric-adapted plant communities the presence of the iconic 'Hawkesbury sandstone' flora offers a rich and unique assemblage of flowering plants. The Mimosoideae contain numerous species of *Acacia* upon which foliage-frequenting, if not foliage-feeding, *Agrilus australasiae, Diphucrania aurocyanea, D. curipennis, D. heroni, D. leucosticta, D. marmorata, D. scabrosula, Melobasis fulgurans, M. gloriosa* and *M. nitidiventris* are to be found. But to find any of these, or their kin, on the masses of *Acacia* flowers that unfurl in spring is a rare thing, for the brush-like flowers of *Acacia*, unlike foliage, apparently offer little or no enticement. By comparison, the zygomorphic, often termed 'egg and bacon', yellow and red-themed flowers of the Faboideae host a distinctive buprestid fauna, especially with respect to the genus *Ethonion* – for example, *Daviesia latifolia* (*Ethonion corpulentum, Diphucrania heroni*); *Dillwynia retorta* (*Ethonion jessicae, E. leai*); *Dillwynia floribunda* (*Ethonion*

---

1   Now considered a synonym of *Temognatha duponti* (Boisudval) (Peterson 2015). Although Carter judged it a rare species *T. duponti* can be common in some seasons (A. Sunholm pers. obs.).

reichei, *E. fissiceps*, *Diphucrania acuducta*); *Jacksonia sco-paria* (*Ethonion reichei*, *Diphucrania acuducta*); *Phyllota phylicoides* (*Melobasis cuprifera*); *Pultenaea ferruginea* (*Ethonion reichei*); *Pultenaea flexilis* (*Ethonion jessicae*); and *Viminaria juncea* (*Diphucrania acuducta*), species of Stigmoderinae, being notably absent.

The regional Casuarinaceae are represented by several species (e.g. *Casuarina cunninghamiana*, *C. glauca*, *Allo-casuarina torulosa*, *A. nana*, *A. distyla*, *A. littoralis*), a few of which are uncommon or highly localised in distribution (e.g. *Allocasuarina glareicola*, *A. portuensis*). The buprestid fauna associated with the foliage of Casuarinaceae is a distinctive one, especially with regard to the genera *Astraeus* (*A. crassus*, *A. dilutipes*, *A. pygmaeus*), *Paracephala* (*P. cyaneipennis*, *P. murina*) and *Germarica* (*G. 'lilliputana'* complex), but also one or two species of *Diphucrania* (most commonly *D. duodecimmaculata*). For the majority of these the association, at least regionally, is essentially an obligate one, though throughout its lowland range *Diphucrania duodecimmaculata* also exhibits a common association with species of *Xanthorrhoea* (Xanthorrhoaceae).

In the shrub–heath, woodland and open forest communities that dominate the region the Myrtaceae are constituted particularly by the mass-flowering tree and shrub genera *Angophora*, *Eucalyptus*, *Corymbia*, *Leptospermum*, *Kunzea*, *Baeckea*, *Melaleuca* and *Calytrix*. A small number of Myrtaceae species (e.g. *Backhousia myrtifolia*, *Syzygium australe*, *S. smithii*, *Syncarpia glomulifera*, *Tristaniopsis laurina*) are associated with the remnants of the once extensive lowland mesic forests of the Illawarra, and scattered rainforest pockets elsewhere, but the buprestid fauna of these communities is relatively poorly studied. However, that of the shrub and woodland communities to be found in Sydney's adjacent national parks, and the Blue Mountains to the immediate west, has historically been the subject of collecting efforts over many decades, the rich fauna to be found there highlighted by H. J. Carter in his book *Gulliver in the Bush* published by Angus and Robertson in 1933. In its pages Carter spoke at length of his buprestid-seeking adventures (sometimes utilising the advantage of riding horseback so as to reach higher flowering branches) among the wealth of flora to be found associated with the Hawkesbury sandstone and Wianamatta shales, noting that more than 100 buprestid species could be found within the Sydney postal district alone. He mentions two plants in particular that are favoured by flower-seeking Buprestidae, the

dwarf apple *Angophora hispida* (previously *Angophora cordifolia*) (Fig. 542), and the tea-tree *Leptospermum polygalifolium* (as *L. flavescens*) (Fig. 543). Carter (1933) cites the words of J. J. Walker (naturalist and once President of the Entomological Society, London): 'I have seen few, if any, more impressive entomological sights than that of a good bush of angophora on a sunny morning with its blossom fairly bending down with the weight of its coleopterous visitors.' *Angophora hispida* is a straggly low tree or shrub in its habit, unfurling masses of cream flowers when in full blossom in December. Its fragrant blossoms produce large volumes of nectar, which in turn attracts a great variety of insects, and in some seasons a rich diversity of nectar-feeding beetles; buprestids among them. The equally perfumed *Leptospermum polygalifolium* can be found associated with stream banks, and flowers profusely in November and December. And again in some seasons it hosts a wealth of insect life. But Carter laments the loss of both species in areas he collected in the 1890s, already in the 1930s given over to urbanisation. Yet these two plant species survive in the national parks with which Sydney is endowed, more than 40 species of Buprestidae being recorded by Williams and Williams (1983) on flowering *Angophora hispida*, and more than 70 species on *Leptospermum polygalifolium*.

Williams and Williams (1983) in their study of the buprestid fauna of the Sydney Basin noted successional flowering at a number of sites by co-occurring myrtaceous *Kunzea ambigua* (Fig. 541), *Leptospermum polygalifolium*, *Angophora hispida* and various *Eucalyptus* species, with *K. ambigua* flowering by late October, *Leptospermum polygalifolium* by mid-November, *Angophora hispida* by late November and *Eucalyptus* species by mid-December. There were, however, within this assemblage varying degrees of overlap in flowering, but within the sequence buprestids noticeably progressed from species to species as the season advanced. They considered that the staggering of flowering might be explained as a strategy by which members of the flowering suite compete for available pollinators present at any one time, and that pollination success within the assemblage might be optimised and competition for pollinators reduced if participating plants staggered their flowering times. The authors further opined that failure of individual plant species to flower in a given season might result in pollinator mortality and loss of fecundity, with any consequent reduction in pollinator numbers impacting on the reproductive potential of those plants flowering later in the sequence. Staggered (intra-specific) flowering of individual plants within populations (e.g. *Angophora hispida*) was also observed in their study, with beetles responding to clines in available energy rewards (nectar), such that large numbers of buprestids could be found visiting a single or a few plants, but be absent or almost so on surrounding individuals of the same plant.

BARRINGTON TOPS

Barrington Tops is a plateau massif, and region of high rainfall, located about 200 km north of Sydney (Figs 537, 538). The main plateau lies at an altitude of 1200–1500 m (rising to 1544 m at Careys Peak), the landscape one of rugged forested slopes that descend steeply on all sides. Its core area is comprised of the Barrington Tops National Park (c. 39 000 ha), this forming the southern-most region of the Gondwana Rainforests World Heritage Area (Adam 1987; Williams 2002; Hunter 2003; Williams 2020a). The region is one of significant phytogeographical interest. Its plant communities are diverse, and include subtropical, warm temperate and cool temperate rainforests; the cool temperate rainforest constitute the southern limit to *Nothofagus moorei* (Fig. 537). In addition there are extensive wet and dry eucalypt-dominated sclerophyll forests, their floristic composition varying with respect to altitude, soil type and aspect. At higher elevations *Eucalyptus pauciflora*, or 'snow gum', forms extensive subalpine woodlands, often in association with open swamps and grasslands.

The region has long drawn scientific interest and investigation. Although not subject to targeted surveys, nevertheless records of its buprestid fauna have accrued as a result of collecting visits undertaken by individual entomologists. Carter (1933) describes his participation in an expedition there in January 1916. The party included various friends and colleagues, the malacologist Charles Hedley and carabid specialist Thomas Sloane among them. They ascended by way of Dungog, and Salisbury on the Williams River, venturing in the last stage by packhorse through forest and 'jungle' to the high plateau. On a later 1925 Sydney University Zoological Expedition Carter participated with a large party that included Professor Lancelot Harrison, lepidopterist Gustavus Waterhouse, Demonstrator in Zoology Lucy Wood, student cartographer Claire Weekes and others, replacing horses with motor vehicles and wagons. Nights were cold, as the 'Tops' in mid-summer frequently are, and the weather stormy.

But Carter collected *Castiarina dimidiata* (Fig. 363) in marshy ground in which patches of *Epacris* grew, and later *Nascioides tillyardi* (Fig. 316) that his friend Robin Tillyard had first discovered at Dorrigo on a previous trip they had taken together. Other buprestids collected included *Castiarina harrisoni*, *Hypostigmodera variegata* (now *Castiarina variegata*) and *Neobuprestis frenchi* (now *N. williamsi*). The blue–purple and yellow *Castiarina variegata* (Figs 460, 461) was first described by the Reverend Blackburn in 1892 from a specimen collected in the Darling Downs, and for many decades was thought restricted to montane and hinterland localities. In more recent years, however, the species has been collected in coastal lowland rainforest (Lansdowne, Iluka) of northern New South Wales. Similarly with *Castiarina harrisoni*, the largest species in the *Castiarina producta* mimicry group (Barker 2006a). First known from hinterland forests of Barrington Tops, and later the elevated forests of Acacia Plateau (far northern New South Wales), the species is also now recorded from lowland and submontane forests on the mid-north coast of New South Wales.

Others followed these first entomologists. R. T. M. Pescott, Director of the then-named National Museum of Victoria, wrote of the 1948 expedition undertaken jointly with staff of the Victorian and New South Wales state museums, these including Alexander Burns and Anthony Musgrave. Their party entered Barrington Tops from the west, venturing to the plateau by car from Scone. Thus they entered a region less well known than the southern region better explored. They needed to ascend '3000 feet in 9 miles' (>900 m to approx. 14 500 m), stopping three times to allow the car radiators to cool, before reaching Tubrabucca on the northern end of the Tops. Writing in *Wildlife* (June, 1948) Pescott gives an overview of the zoological encounters and expedition outcomes, and describes the proliferation of insects on flowering *Leptospermum* (Fig. 538). He mentions that 15 species of Buprestidae were collected, though only giving the large yellow and black *Stigmodera macularia* (Figs 182, 469) by name. It is a common species, reliably to be found each year, and also occurring in Queensland and Victoria, but surprisingly absent from Tasmania (Bellamy 2002).

As currently understood the buprestid fauna comprises a suite of species associated with eucalypt-dominated sclerophyll forest and woodland, and a smaller group that is associated with rainforest. The former comprises species characteristic of elevated open forests extending from at least the Blue Mountains in the south, north to the New

England Tablelands, within which there appears to be a degree of 'latitudinal sifting'. Most records have been obtained by collecting from flowering *Leptospermum polygalifolium*, a widespread, large shrub or small tree occurring on forest margins but also commonly as dense thickets along stream banks and drainage lines. Its white flowers are prolific especially from mid-December to mid-January and later. In addition to the species of *Castiarina* earlier mentioned, species include *C. bifasciata*, *C. bremei*, *C. brutella*, *C. cydista*, *C. discoflava*, *C. hilaris*, *C. nasuta*, *C. producta*, *C. watkinsi*, *C. zecki*, and the distinctive *C. eborica*, a species restricted throughout its range to elevated forests. *Agrilus hypoleuca* is recorded from *Acacia* foliage in open forest. The fauna recorded from rainforest, owing to the focus of collecting effort, is essentially that of cool temperate rainforest, the few species recorded being reared from dead branches of *Nothofagus moorei*; *Nascioides tillyardi*, *N. macalpinei*, *Melobasis ignipicta* and *M. hypocrita*. Although there are numerous mass-flowering canopy tree species (Adam 1987; Hunter 2003) these present problems of accessing their high blossoms and so the rainforest flower-frequenting fauna is poorly known.

NEW ENGLAND TABLELANDS AND ASSOCIATED MONTANE RAINFORESTS

Stretching north from the Barrington Tops to the Queensland border, and into that state's south-east corner, is a mountainous realm of great biogeographical interest, a biodiversity 'hot spot', where plants and animals of diverse evolutionary lineages intermingle and overlap in complex ecological assemblages and with many relic forms surviving in isolation (Adam 1987; Williams 2002; Hunter 2003; Williams 2020a, 2021). Occupying this region is a great expanse of elevated eucalypt forests and woodland interspersed with a diversity of rainforest formations (Figs 531, 532) that have survived episodes of fire, invasive species, clearing for farmland, and the planting of forestry monocultures. As a consequence, continuity of native vegetation is often interrupted and tenuous, the corridors available for the movement of fauna and the dispersal of plant genetic material from locality to locality irregular and uncertain. Nevertheless, many of the plant communities are now within the Gondwana Rainforests World Heritage Area (some 366 507 ha) that extends patchwork-like into south-eastern Queensland, with focal reserves in New South Wales in sequence from the Barrington Tops

northwards largely comprising Werrikimbe, Oxley Wild Rivers, New England, Dorrigo, Washpool, Gibraltar Ranges, Nightcap, Border Ranges and Mt Warning national parks, and Mt Seaview and Mt Hyland nature reserves (in New South Wales), and reserves of the Main Range, Mount Barney, Mount Chinghee, MacPherson, Lamington and Springbrook World Heritage scenic rim in south-east Queensland. Within these parks are mountain peaks and prominences of significant height and visual impact (e.g. Mount Barney 1359 m, Mt Lindsay 1177 m, Mt Warning 1156 m, Point Lookout 1563 m, Mt Banda Banda 1258 m). Associated with the world heritage reserves are areas of state forest that reside outside of the World Heritage Area boundary, as well as a mosaic of unreserved, privately owned forest.

Of all these some, such as Mount Tambourine and Dorrigo National Park, are well trodden – casual or fleeting as the individual moments of entomological investigation often have been. Yet much remains unstudied and the outcomes of the earliest visits at best are poorly documented. In the 1940s the area about Dorrigo was the haunt of W. Heron, who sent diverse specimens south to the Australian Museum at Sydney. Heron was an enigmatic forestry worker and naturalist after whom Carter named *Diphucrania heroni* (Fig. 232). Further north at Acacia Plateau, located on the New South Wales–Queensland border, Camille Deuquet (active in the 1930s–60s) discovered *Castiarina acuta*, and a little to the west at Stanthorpe, he found *Castiarina rutila* and *C. harslettae*; the latter named by him after Jean Harslett, daughter of Alex Gemmell (farmer and entomologist who resided at Glen Aplin), for whom he had named *Temognatha gemmelli* (Figs 193, 478). Carter also ventured to the region. In 1910, in company with T. G. Sloane, Carter made a hurried visit to 'the Dorrigo', travelling up from Belligen on the coast via mail coach to the little hamlet of Never Never. A picnic area in Dorrigo National Park now preserves this name. It was July, it was cold, some snow fell, and it was not a time of the year for finding buprestids. Instead they 'rolled logs' and collected ground beetles. Later in the same year the two visited Acacia Creek. Carter described their visit as one timed in the best season for collecting, but when writing of their exploits many years later he reflected that he had managed to mislay his records of the Acacia Creek catch. In 1913, with Arthur Lea, and in 1914 once more with Sloane, Carter again ventured north, on both occasions to Tambourine Mountain (he did not like calling it 'Mount Tambourine') close to the coast; but the mountain was detached somewhat from the MacPherson Range. The eminent Robin Tillyard had walked its 'jungle' forests before him. But Tillyard was interested in dragonflies, not beetles (though he collected these as an 'aside'). Rowland Illidge, naturalist and pioneering entomologist, and Henry Hacker, entomologist with the Queensland Museum, also were visitors to Tambourine Mountain, as well as to the high mountain rainforests and eucalypt forests of the main MacPherson Range further inland. Both have buprestids named in their memory (*Melobasis illidgei*, *Trachys hackeri*, *Castiarina hackeri*). In the MacPherson Range Hacker collected what was to become the holotype of the still poorly known *Nascioides nulgarra*, a species reminiscent of *Nascioides tillyardi* found further to the south, but with markedly reduced red elytral spots, yet also utilising *Nothofagus moorei* as a larval host: in the Border–MacPherson ranges *Nothofagus* reaches its northern-most distribution. From the same locality Hacker also collected specimens of what were to be later named *Nascioides bicolor* (otherwise known from Samford, Nightcap Range and Dorrigo National Park) and *N. falsomultesimus* (restricted to the montane forests of northern New South Wales and south-eastern Queensland) (Williams 1987).

The region is very rich in *Castiarina* species of the 'producta' mimicry complex (e.g. *C. acuminata*, *C. beatrix*, *C. delicatula*, *C. delta*, *C. jacki*, *C. kitchini*, *C. obliqua*, *C. producta*, *C. pseudasilida*, *C. spectabilis*, *C. zecki*), and although no study has attempted to survey and table the broader diversity buprestid fauna of individual reserves numerous examples of specific localities can be found in Barker (2006a), Curletti (2001), Levey (2012), Levey and Bellamy (2013), Williams (1987, 2002, 2020a) and Williams and Bellamy (2002). As with Barrington Tops already mentioned, the buprestid fauna consists of one associated with rainforests (and their often adjacent wet sclerophyll forest) and a second suite characteristic of open forests, these interspersed with shrub and heath complexes. However, the adult fauna that frequents flowers can be 'fluid' in their movements across plant boundaries, for individuals respond to the availability of floral resources and can undertake flight patterns in response to the availability of flowers beyond that of the habitat of their larval hosts, a caveat being that these need to have open, readily accessible flower structures (Williams 2021). Peak time for the emergence of flower-frequenting buprestids is a little later than that on the coast, with flowering events (particularly of Myrtaceae) most intense in late December to January. However, many beetles frequent the highest blossoms such that they often avoid even the longest of extendable nets.

The foliage- and branch-frequenting rainforest fauna includes *Agrilus deauratus*, *A. carterellus*, *Aaaaba fossicollis*, *A. nodosus*, (Agrilinae), *Anilara obscura*, *A. olivia*, *Melobasis hypocrita*, *Nascioides bicolor*, *N. falsomultesimus*, *N. macalpinei*, *N. multesimus*, *N. nulgarra*, *N. pulcher*, *N. tillyardi*, *N. viridis*, *Neobuprestis williamsi*, *Notographus sulcipennis*, *Pseudanilara cupripes* (Buprestinae), *Paracupta lambertii* (Chrysochroinae) and *Prospheres aurantiopictus* (Polycestinae). These are best collected by beating foliage but especially by rearing from timber billets collected from fallen branches. The flower-frequenting species include *Calodema regale*, *Castiarina acuta*, *C. ariel*, *C. beatrix*, *C. harrisoni*, *C. hilleri*, *C. liliputana*, *C. warningensis*, *Metaxymorpha grayii* and *M. imitator* (Stigmoderinae).

Records for the 'open forest' fauna are more diverse, reflecting the generally greater ease of collecting the fauna, and consequent records accruing over multiple visits, in the region inland from Armidale and Glen Innes, and north to the Queensland border, hence Stanthorpe being especially well-collected. However, the records for flower-frequenting stigmoderine species (largely collected from *Leptospermum*) likely include species that also frequent the high canopies of mass-flowering rainforest trees. Records of particular note from sclerophyll forests, and shrub and heath complexes include *Agrilus armstrongi*, *Diphucrania heroni*, *Synechocera burnsi* (Agrilinae), *Melobasis cupreovittata*, *M. nervosa*, *M. simplex*, *M. williamsi*, *Nascio simillima*, *Nascioides carissimus*, *N. olliffi*, *N. parryi*, *Pseudanilara purpureicollis* (Buprestinae), *Astraeus crassus*, *A. cyaneous*, *A. jansoni* (Polycestinae), *Castiarina alecgemmelli*, *C. armata*, *C. binotata*, *C. brutella*, *C. costalis*, *C. harslettae*, *C. inflata*, *C. pseudoerythroptera*, *C. rayclarkei*, *C. rutila*, *Temognatha gemellii* and *T. maculiventris* (Stigmoderinae).

LITTORAL RAINFORESTS OF NORTHERN NEW SOUTH WALES

Littoral rainforests comprise closed forest formations found on beach sand dunes, coastal headlands and some estuary associated islands (Adam 1992) (Fig. 533). They extend from north-west Western Australia, along the Queensland and New South Wales coastlines, and to eastern Victoria. One site, Iluka Nature Reserve on the New South Wales far north coast, is included within the

Gondwana Rainforests World Heritage Area (Adam 1987), the only littoral rainforest to be so reserved. Southern formations in New South Wales and Victoria are floristically depauperate; however, headland formations tend to be more floristically complex (e.g. Sea Acres Nature Reserve, Hallidays Point), regionally variable in composition, and exhibit strong relationships to regional rainforests further inland. They represent distinctive habitat-restricted communities that have been subject to extensive clearing, fragmentation, fire impacts and weed invasion since European occupation, yet the composition of their invertebrate fauna has been poorly studied. In 1912 H. J. Carter visited Brunswick Heads and in 1925 he visited Port Macquarie (Carter 1933). Examples of littoral rainforest are to be found at both locations, but he makes no mention of the buprestid fauna that might occur there.

Here we briefly consider what is known of the buprestid fauna of littoral rainforests that survive on the coastline of northern New South Wales. Many species recorded from these remnants have also been recorded from hinterland rainforests (Williams 1993, 1995, 2002; Williams and Adam 2019). Records of phytophagous taxa have been obtained by visual observation of species in flight, sweeping foliage or by rearing from timber billets cut from fallen branches or dead stems.

Of the phytophagous fauna the largest species are *Cyrioides australis* and *Paracupta lambertii*. Adult *Cyrioides australis* can be found resting on the leaves and branches of coast banksia, *Banksia integrifolia*, a large tree associated with regenerating littoral forests and littoral rainforest margins. *Cyrioides australis* can also be encountered in dry sclerophyll forest further inland. *Paracupta lambertii* is a rainforest specialist (e.g. Harrington, Iluka), being also recorded from hinterland subtropical rainforests (Lansdowne, Bowraville) but absent from open forests. In littoral rainforest it has been collected from the foliage of the tree *Elaeocarpus obovatus*. *Cyrioides australis* and *Paracupta lambertii* have been observed flying across open spaces during hot summer days. Additional foliage-associated species include the diminutive *Helferella gothmogoides* and *H. manningensis* (Williams and Weir 1987). Their larval and adult hosts remain unknown, and to date they have only been collected from littoral rainforest remnants (Harrington, Manning Point, Camden Haven) by general sweeping of living foliage or from intercept traps. Similarly with tiny individuals of *Aphanisticus*, which have only been recorded so far from flight intersection traps set on forest

margins and below canopy openings, but from which no host data can be obtained. Adults of the small and ovoid *Paratrachys australis*, a species also recorded from floodplain subtropical rainforest remnants, frequent the vine *Trophis scandens* but its larval host remains elusive (though possibly species of *Ficus* [Bellamy and Williams 1995]). Equally so with a species of *Habroloma*, which has been swept from the foliage of trees growing on forest margins, but like *Paratrachys* its larval food plant is also yet to be discovered. The host relationships of *Maoraxia auroimpressa* are comparatively well known. In late spring and summer the adults frequent the sunlit leaves of numerous tree species and utilise a diversity of rough barked trees as larval hosts. These include *Acronychia* sp., *Podocarpus elatus*, *Elaeodendron australe*, *Alectryon coriaceus* and *Guoia semiglauca* (Appendix 2; Bellamy et al. 2013). *Melobasis splendida* has been reared from dead canes of the rampant vine *Maclura cochinchinensis* and branches of *Elaeodendron australe*. Several species of *Agrilus* inhabit littoral rainforests. Adults may be swept or beaten from foliage, but *Agrilus carterellus* has been reared from *Claoxylon australe*, and *A. deauratus* and *A. frenchi* from *Drypetes deplanchei*. *Pseudanilara* is represented by *P. cupripes*, a widespread species in regional rainforest and wet sclerophyll forest. Adults frequent foliage but occasionally are to be found on flowers of mass-flowering canopy trees. They are very active in sunlight, and can be found ovipositing on fallen branches. Also attracted to fallen branches as ovipositing sites is *Chrysobothris simplicifrons*. This is a species recorded from a variety of regional rainforest formations. Its known larval host plants include the tree *Alphitonia excelsa* and the large shrub–small tree *Myrsine variabilis*. Adults are very keen-eyed and quickly move out of sight when approached.

The flower-frequenting Buprestidae have been extensively investigated, though the published records are based almost wholly on studies undertaken in the Manning Valley (e.g. Williams 1995; Williams and Adam 2019). The littoral rainforest fauna appears to be an impoverished one, relative to the rich fauna recorded from flowering trees and shrubs in submontane and floodplain rainforests to the near-west (Williams 1993, 1995, 2002; Williams and Adam 2019). Of *Castiarina*, usually a prolific genus, Williams (1995) recorded only *C. cydista* (on *Scolopia braunii*), *C. decemmaculata* (*Syzygium smithii*), *C. acuminata* (*Alphitonia excelsa, Guoia semiglauca*), *C. neglecta* (*Alphitonia excelsa*), and *C. producta* (*Euroschinus falcatus*). From the same study can be added *Torresita cuprifera* and *Neocuris* sp. (on *Alphitonia excelsa*). However, years of further observation at the same sites have added only one additional flower-frequenting adult food plant record, that being *Castiarina acuminata* on *Acronychia imperforata*. A record of particular interest is that of *Castiarina variegata*, but this from the Iluka Nature Reserve world heritage site further to the north.

ALPINE AND MONTANE SOUTHERN NEW SOUTH WALES AND EASTERN VICTORIA

This vegetation-varied 'snow-country' region includes extensive areas of high altitude reserves such as Brindabella (approx. 18 400 ha), Kosciuszko (600 000 ha), and Namadgi (106 000 ha) national parks in New South Wales and the Australian Capital Territory respectively, and the Alpine (approx. 640 000 ha) and Mount Buffalo (31 000 ha) national parks in Victoria (Figs 549, 550). Commonly referred to as the 'Australian Alps' and 'Snowy Mountains' the region's elevated peaks include Mount Gudgenby (1740 m, Namadgi NP), The Horn (1723 m, Mt Buffalo NP), Mt Feathertop (1922 m, Alpine NP), and Mt Kosciuszko (2228 m, Kosciuszko NP) – this the highest peak in Australia, upon which late-lasting snow patches may persist for many months.

Relative proximity to the sea has a moderating influence on summer as well as winter temperatures with average summer temperatures in the Kosciuszko area ranging from about 14 to 16° Celsius but with increasing climate change-induced temperatures threatening unique fauna and flora throughout alpine and subalpine zones. Truly alpine zones occur above 1800 m, with subalpine zones existing down to about 300–500 m below the treeline (Costin *et al.* 2000). The flora includes a diversity of alpine herb fields, heaths and tussock grassland, and in subalpine areas *Eucalyptus pauciflora* 'snow gum' woodland[2] (from about 1500 m extending up to about 1800 m) and *Carex–Sphagnum–Epacris* fens and bogs (Costin *et al.* 2000; Doherty *et al.* 2015). Extensive representations of eucalypt-dominated forest and woodland occur at lower elevations, and in eastern Victoria there are spatial and floristically significant examples of both lowland and elevated temperate rainforest (e.g. in the Strzelecki Ranges, Wilsons Promontory, Central Highlands, East Gippsland)

(Cameron 1987; Adam 1992). Often there is a rainforest understorey developed below a eucalypt canopy. In the East Gippsland region occurs the Errinundra Plateau (Fig. 551) with unique, though spatially restricted, cool temperate moss thickets and closed scrubs of variable height characterised by endemic forms of *Podocarpus* and *Tasmannia*, but with pure stands of cool temperate 'sassafras'-termed rainforest co-dominated or dominated by *Atherosperma moschatum*, *Elaeocarpus holopetalus* and *Acacia melanoxylon* (Cameron 1987). However, here there is an apparent anomalous absence of *Nothofagus cunninghamii*, a dominant canopy tree otherwise to be found in rainforest communities of the Otway Ranges (south-western Victoria [Fig. 554]), the Victorian Central Highlands, Strzelecki Ranges and Wilsons Promontory, as well as in Tasmania. Its absence at Errinundra precludes the occurrence there of the obligate *Nothofagus cunninghamii*-associated *Nascioides quadrinotatus*. This buprestid species occurs in Tasmania, and in Victoria has been collected at lower elevations (~800 m) at Tanjil Bren (Yarra Ranges National Park north-east of Melbourne) from *Nothofagus cunninghamii* (Williams 1987; Williams and Bellamy 2002).

Early explorers to the region included the Polish-born Lhotsky (1834) and Strzelecki (credited with the first ascent of Mt Kosciuszko in 1840) (Costin *et al.* 2000). During the 1850s the German-born botanist Ferdinand von Mueller undertook several expeditions to the alpine regions of eastern Victoria and the Kosciuszko area. Later botanist visitors included J. H. Maiden, I. Luckie and H. K. Petrie. The German-born entomologist Richard Helms visited the region in 1889 and again in 1893. The results of his collecting were published in 1890 (in the *Records of the Australian Museum*) and 1896/1897 in the *Journal of the Royal Geographical Society of Australasia*. H. J. Carter visited Jindabyne, Kosciuszko and Thredbo a number of times during the early 1900s. In December of 1905 he found *Eucalyptus viminalis* 'in full flower and a fine lot of Buprestidae and other flower-haunting species were taken' (Carter 1933). Other Coleoptera did not escape his attention, nor did a suite of *Castiarina* found associated with flowering *Leptospermum*. In 1926 Carter made his final visit, this time in the company of Alexander Nicholson, Lecturer in Entomology at the University of Sydney and later Chief of the CSIR/CSIRO Division of Economic Entomology. The season was hot and dry but a new *Castiarina* (*C. flavoviridis* [Fig. 138]) was discovered on *Leptospermum* bushes.

---

2   Cerambycid-induced and climate-associated stress are responsible for extensive dieback of *Eucalyptus pauciflora* in alpine and subalpine zones.

A foundation work tabulating the Victorian fauna was undertaken by Burns and Burns (1992). This forms the basis for species listed in Table 4, but with additional reference to the numerous papers by Barker, Bellamy, Carter, Curletti and Levey previously cited. The alpine herb fields are depauperate with respect to buprestids but at lower elevations (e.g. in the vicinity of Thredbo, Island Bend) summer-flowering *Leptospermum* shrubs support a diversity of *Castiarina*. Species such as *Castiarina australasiae, C. bifasciata, C. delectabilis, C. dimidiata, C. erythromelas, C. flavopicta, C. hilaris* and *C. sexplagiata* are widely distributed forms (Bellamy 2002; Barker 2006a) but *C. alpestris, C. erasma, C. helmsi, C. insularis, C. flavopurpurea* and *C. flavoviridis* are more distinctive. Of these *Castiarina helmsi* and *C. flavoviridis* were described by Carter (1906, 1927) from material collected by him at or in the vicinity of Mt Kosciuszko and that of *C. flavopurpurea* (Fig. 137) from specimens emanating from the nearby Jindabyne, as well as specimens from the Blue Mountains to the north (Carter 1908).

The general habitus of *Castiarina flavopurpurea* and *C. flavoviridis* bring to mind that of the endemic Tasmanian species *C. insculpta, C. macquillani, C. tasmaniensis* and *C. virginea*, this group possibly representing an ancient temperate lineage within the genus.

In addition to those species cited above, the Stigmoderinae include numerous species from regional montane forests, woodlands and associated shrub and heath communities (Table 4). *Stigmodera* is represented by the widespread *S. macularia*, and there are several species of *Temognatha* (Table 4). Many of the recorded *Castiarina* also occur at lower elevations (e.g. *Castiarina scalaris, C. sexplagiata, C. supergrata, C. thomsoni*) and in more xeric-adapted plant communities (e.g. *C. decemmaculata, C. delectabilis, C. flavopicta, C. rufipennis*). Also known from the region, though rarely encountered, is *Castiarina interstincta*. This species was originally named and described by Barker (1983; see also 1990, 2006a) as *Stigmodera (Castiarina) variegata*, and later cited by Burns and Burns (1992) as *Castiarina variegata*; a cause for confusion with *Castiarina variegata*, this distinct second

Table 4. Examples (given alphabetically) of buprestid species from alpine and montane southern New South Wales and eastern Victoria.

Agrilus aurovittatus, Agrilus hypoleucus
Astraeus jansoni, Astraeus mastersi
Castiarina alpestris, Castiarina attenuata, Castiarina australasiae, Castiarina bella, Castiarina bifasciata, Castiarina bremei, Castiarina burnsi, Castiarina crenata, Castiarina cruentata, Castiarina delectabilis, Castiarina deyrollei, Castiarina dimidiata, Castiarina erasma, Castiarina erythromelas, Castiarina erythroptera, Castiarina flavopicta, Castiarina flavopurpurea, Castiarina gordonburnsi, Castiarina helmsi, Castiarina hilaris, Castiarina imitator, Castiarina interstincta, Castiarina interstitialis, Castiarina kershawi, Castiarina livida, Castiarina media, Castiarina montigena, Castiarina nasuta, Castiarina octomaculata, Castiarina octospilota, Castiarina parallela, Castiarina punctatosulcata, Castiarina rectifasciata, Castiarina rufipennis, Castiarina scalaris, Castiarina sexguttata, Castiarina sexplagiata, Castiarina spinolae, Castiarina supergrata, Castiarina thomsoni, Castiarina wilsoni, Castiarina xanthopilosa, Castiarina xystra
Dinocephalia thoracica
Diphucrania acuducta, Diphucrania cupreola, D. duodecimmaculata, Diphucrania leucosticata, Diphucrania marmorata
Iridotaenia albivittis
Nascio vetusta
Nascioides parryi
Neobuprestis frenchi
Neocuris pauperata
Pachycisseis bicolor
Pseudanilara purpureicollis
Melobasis innocua, Melobasis nervosa, Melobasis picticollis, Melobasis purpurascens, Melobasis sordida, Melobasis viridisterna
Selagis aurifera, Selagis caloptera
Stigmodera macularia
Torresita cuprifera
Temognatha maculiventris, Temognatha menalcas, Temognatha mitchellii

species first named and described by the Reverend Canon Thomas Blackburn (1892) as *Hypostigmodera variegata*.

Buprestinae recorded from montane elevations include *Nascioides parryi* and the related *Nascio vetusta*, both species widely distributed in the open forests of south-eastern Australia. Adults may be encountered active on sunlit eucalypt trunks. Additional buprestine genera include foliage-associated *Melobasis* and flower-frequenting *Selagis caloptera*, *S. aurifera* and *Torresita cuprifera*. Recorded species of Polycestinae (*Astraeus*) and Agrilinae (*Agrilus*, *Dinocephalia*, *Diphucrania*, *Pachycisseis*) are generally those with wide latitudinal and altitudinal distributions in south-eastern Australia.

NORTH-WESTERN VICTORIA

North-western Victoria differs greatly from the vegetation communities of the state's east. It is a region characterised by a semi-arid steppe-like climate, and includes a number of Victoria's oldest and largest national parks, such as Wyperfield (357 000 ha), Little Desert (132 600 ha), Big Desert Wilderness Park (141 700 ha) and the Murray-Sunset National Park (633 000 ha) (Figs 552, 553). However, many reserves are situated in otherwise heavily cleared agricultural landscapes such that they are effectively habitat 'islands'.

The reserve system of north-western Victoria conserves extensive complex of mallee, heathland, native grasslands, dune associated low shrubland, *Callitris* pine (Cupressaceae) and 'she oak' (Casuarinaceae) woodlands and, in the case of the Murray-Sunset National Park, semi-arid woodlands. 'Mallee'-dominated landscapes encompass a number of distinct plant assemblages. The so-named 'deep sand mallee' of the Sunset district is dominated by *Eucalyptus costata* with *Acacia ligulata* and *Leptospermum coriaceum* as characteristic associates. In the 'loamy sand mallee' of the Big Desert-Wyperfield National Park are to be found *Eucalyptus leptophylla* and *E. costata* as dominant eucalypts with common associate species being *Bertya mitchelli*, *Callitris verrucosa* and *Glischrocaryon behrii*. In the expansive 'mallee–heath' communities of the Big Desert occur *Eucalyptus leptophylla* and/or *E. costata* as dominants, with shrub associates that include *Allocasuarina pusilla*, *Aotus subspinosus*, *Hysterobaeckia behrii*, *Baeckia crassifolia*, *Banksia marginata*, *Callitris verrucosa*, *Calytrix tetragona*, *Comesperma calymega*, *Glischrocaryon behrii*, *Grevillea ilicifolia*, *Hakea muelleriana*, *Leptospermum coriaceum*,

Leucopogon rufus and *Melaleuca wilsonii*. Species dominant or co-dominant in the intermediate mallee–heath and loamy sand mallee communities of the Big Desert include *Acacia spinescens*, *Callitris verucosa*, *Calytrix tetragona*, *Eucalyptus costata*, *E. leptophylla* and *Leptospermum coriaceum*. An overview of the region's plant communities can be found at the Royal Botanic Gardens Victoria web portal (https://vicflora.rbg.vic.gov.au/pages/bioregions).

A major landform feature of north-western Victoria is an extensive low consolidated sandy dune system (<200 m in altitude) of Pleistocene age (Fig. 552). The panorama formed extends into the adjacent South Australia, and gives the region a heightened sense of 'wildness' and consequent visitor attraction. But the dunes also create a phenomenon of shelter for buprestids, for these in season gather on low flowering shrubs on the leeward side when winds blow across the vast vegetated landscape.

However, the region is one of low endemism (Levey 2012) with many of the buprestids recorded from north-western Victoria occurring elsewhere in that state (Burns and Burns 1992; Barker 2006a), and species otherwise also occurring in South Australia, Western Australia, New South Wales and/or Queensland (Bellamy 2002; Barker 2006a; Levey 2012, 2018; Lang 2020). Nevertheless, north-western Victoria is an interesting bioregion in as much as it represents the eastern-most range of species that are otherwise Western Australian–South Australian in their distribution (see Table 7), it illustrates the east–west continuum of species ranges, and provides examples of semi-arid zone associated fauna (see Table 5).

The region is one of historical interest to entomologists and collectors alike. Consequently, the buprestid fauna, especially that associated with flowering Myrtaceae (e.g. *Baeckea*, *Calytrix*, *Eucalyptus*, *Leptospermum*), has been extensively investigated (examples in Table 5); in recent decades drawing an increasing interest, but also with earlier collectors including Charles Oke, J. E. Dixon, John Goudie, Keith Hateley, and Gordon and Joy Burns. Starting in 1909 John Goudie published a series of papers in the *Victorian Naturalist* on the beetle fauna of north-western Victoria, the issue of 1920 noting the habits of various species of *Anilara*, *Astraeus*, *Bubastes*, *Castiarina*, *Diadoxus*, *Melobasis*, *Temognatha* and others. Keith Hateley was a naturalist and resident of the small town of Kiata in western Victoria, advocated the conservation of native habitat in that region, and later became the first

Table 5.   Examples (given alphabetically) of buprestid species from North-western Victoria (records mainly from the Big and Little Desert areas).

Agrilus australasiae
Anilara adelaidae, Anilara longicollis
Astraeus badeni, Astraeus major, Astraeus irregularis
Bubastes barkeri, Bubastes burnsi, Bubastes globicollis, Bubastes vagans
Calotemognatha yarrelli
Castiarina adelaidae, Castiarina aeneicornis, Castiarina argillacea, Castiarina aurantiaca, Castiarina burnsi, Castiarina castelnaudii, Castiarina crux, Castiarina decemmaculata, Castiarina distinguenda, Castiarina dugganensis, Castiarina flavopicta, Castiarina flavosignata, Castiarina fulviventris, Castiarina gardnerae, Castiarina gibbicollis, Castiarina goudiana, Castiarina hateleyi, Castiarina ignea, Castiarina jekelli, Castiarina jospilota, Castiarina kiatae, Castiarina kirbyi, Castiarina leai, Castiarina marginata, Castiarina marginicollis, Castiarina media, Castiarina obscura, Castiarina ochreiventris, Castiarina octomaculata, Castiarina ovata, Castiarina pallidipennis, Castiarina pallidiventris, Castiarina parallela, Castiarina perlonga, Castiarina piliventris, Castiarina punctiventris, Castiarina recta, Castiarina robusta, Castiarina rufa, Castiarina simulata, Castiarina straminea, Castiarina subpura, Castiarina subtincta, Castiarina triramosa, Castiarina uncata, Castiarina varidolinea, Castiarina vittata, Castiarina xanthospilota
Chalcophorotaenia exilis
Chrysobothris mastersii
Diadoxus erythrurus, Diadoxus jungi, Diadoxus regius
Dinocephalia burnsi
Diphucrania aenigma, Diphucrania constricta, Diphucrania fulgidicollis, Diphucrania minutissima, Diphucrania rubicunda, Diphucrania vicina
Ethonion leai, Ethonion maculatum
Germarica lilliputana
Hypocisseis minuta, Hypocisseis suturalis
Julodimorpha bakewelli
Melobasis abnormis, Melobasis andersoni, Melobasis angustecostata, Melobasis cupreovittata cupreovittata, Melobasis fulgurans, Melobasis gratiosimma, Melobasis lathami, Melobasis nervosa, Melobasis nobilitata, Melobasis propinqua verna, Melobasis pyritosa, Melobasis sordida, Melobasis splendida, Melobasis thoracica
Merimna atrata
Microcastalia globithorax
Neobubastes aureocincta
Neobuprestsis peroni
Neocuris asperipennis, Neocuris dichroa, Neocuris discoflava, Neocuris doddi, Neocuris fairmairei, Neocuris guerinii, Neocuris monochroma, Neocuris viridimicans
Neospades chrysopygia
Paracephala murina, Paracephala pistacina
Pseudanilara bicolor, Pseudanilara cupripes, Pseudanilara pilosa
Selagis aurifera, Selagis viridicyanea
Stanwatkinsius lindi
Stigmodera macularia, Stigmodera sanguinosa
Temognatha congener, Temognatha duboulayi, Temognatha flavicollis, Temognatha flavocincta, Temognatha flavomarginata, Temognatha flavicollis, Temognatha fortnumii, Temognatha heros, Temognatha, maculiventris, Temognatha pubicollis, Temognatha sanguiniventris, Temognatha tricolorata, Temognatha vitticollis

ranger of the Little Desert National Park (Robin 1994). He collected a number of new and poorly known species (e.g. *Castiarina hateleyi, C. kiatae, C. uncata*), many of which were deposited in the collection of the South Australian Museum. Gordon and Joy Burns travelled and collected widely in Australia, but particularly so in their home state Victoria; this described in Barker (2006a). In addition to long hours assisting in the curation of the Museum of Victoria's buprestid collection, they contributed to the Entomological Society of Victoria's ENTRECS project,

'The distribution of Victorian jewel beetles (Coleoptera: Buprestidae) – an ENTRECS project'. This was a checklist of the Buprestidae of Victoria, the distribution of each species (to 1990) assigned to 'blocks' defined alphabetically by units of latitude and longitude – a mammoth task and testimony to their undertaking. The project outcome was published as an Occasional Paper of the Museum of Victoria (Burns and Burns 1992).

SOUTH AUSTRALIA

Much of the state receives less than 750 mm annually, with native vegetation being dominated by woodland (Figs 555–558). As a general overview, the vegetation of South Australia is composed of open forests in the south-east (e.g. Mount Lofty Ranges, southern Flinders Ranges, western Kangaroo Island, vicinity of Port Lincoln), extensive woodland dominated by eucalypts and occasional *Callitris*, Casuarinaceae, *Melaleuca* and *Acacia*, multi-stemmed 'mallee', open and closed heaths and shrublands, tussock grasslands, and coastal dune and saltmarsh complexes (Specht 1972). These support a diversity of buprestid species (Tables 5, 6). But there has been significant clearing of native vegetation for agriculture, especially in the south-east and the Eyre and Yorke peninsulas. The impact of such habitat loss on the fauna is uncertain, but Tepper, writing in *Common Insects of South Australia* in 1887, already notes the rarity of some species and the reduction in abundance from past years (e.g. *Temognatha congener*, *T. flavomarginata*, *T. flavocincta*, *T. fortnumii*).

South Australia has been fortunate in having a number of eminent coleopterists in residence and, though the focus of some was on families other than Buprestidae, collectively their attention has contributed significantly to our understanding of the fauna as it now exists. In years past the Government Entomologist Arthur Mills Lea and Canon Thomas Blackburn amassed substantial general Coleoptera collections, that of Blackburn being purchased by the Natural History Museum, London. In more recent years Shelley Barker (University of South Australia) and Eric Matthews (South Australian Museum) have left a great research legacy. Shelley Barker undertook the herculean task of revising not only *Castiarina* but also *Astraeus* and *Diphucrania*, with many of his types now deposited in the South Australian Museum, Adelaide. In addition to producing the numerous volumes of the *Beetles of South Australia*, in which Buprestidae are included in Volume 4, Eric Matthews confronted the Australian

Scarabaeinae and Tenebrionidae, his studies resulting in the identification of major centres of endemism for the Australian beetle fauna (Matthews 1972; Matthews and Bouchard 2008). South Australia is also fortunate in that 'whole of fauna' identification guides are available. In addition to Matthew's (1985) key to genera Lang (2020) provides an extensive and unique visual portal to species and host associations.

As with that of north-western Victoria, the buprestid fauna is essentially one derived from south-west and south-eastern Australia (see Tables 5, 6), and is considered to be one of generally low endemism (Lang 2020; Levey 2012). Levey suggests that, at least for *Melobasis*, the low level of endemism 'is likely to be a reflection of the drier climate of the region during some parts of the Pleistocene when conditions would have been unsuitable for the survival of species not adapted to arid conditions'.

Nevertheless, more than 320 species are recorded (Lang 2020), the majority of described species composed of *Castiarina* (c. 116 spp.), *Melobasis* (c. 40 spp.), *Diphucrania* (27 spp.), *Temognatha* (23 spp.) and *Neocuris* (19 spp.), and with substantial numbers of *Astraeus* (9 spp.), *Bubastes* (8 spp.) and *Agrilus* (7 spp.) (Table 6). There are also numerous species assigned to *Anilara* (15 in Lang 2020) but this is a genus requiring critical revision.

The fauna can be grouped into: (1) species distributed from Western Australia to South Australia; (2) species distributed from Western Australia to north-western Victoria; (3) species restricted to South Australia and Victoria; and (4) more widespread species also occurring in New South Wales, Queensland and/or Tasmania and occasionally recorded from the Northern Territory. Selected examples from groups 1–3 are given in Table 7.

There are a few apparent endemics. Species considered restricted to South Australia include *Bubastes deserta* (a poorly known species not recently collected), *Castiarina creta*, *C. eyrensis*, *C. guttata*, *C. julia*, *C. keyzeri*, *C. lukini*, *C. melrosensis*, *C. nota*, *C. tepperi*, *C. vanderwoudeae*, and *Diphucrania adusta*, *D. chalcophora*, *D. robertfisheri* and *Selagis obscura*. *Astraeus* includes a unique species, *A. aridus*, the type from Puttapa Gap, near Copley, collected from the leaves and stems of *Melaleuca glomerata* (Barker 1989). *Castiarina uptoni* represents an additional arid/semi-arid zone species. There are a number of smaller genera with representatives whose recorded range is restricted to Western Australia and South Australia (e.g. *Euryspilus australis*, *E. chalcodes*, *Microcastalia scintillans*, *Notobubastes costatus*), or Western Australia and South

Table 6.   Genera and approximate number of species* recorded from South Australia (after Barker 2006a, 2006b, 2006c; Bellamy 2002; Lang 2020; Levey 2012, 2018)
*Species values may be approximate.

Genus	spp.	Genus	spp.
Agrilus	7	*Melobasis*	40
Anilara	15	*Merimna*	1
Astraeus	9	*Microcastalia*	2
Australorhipis	1	*Nascio*	1
Bubastes	8	*Nascioides*	1
Buprestis/Cypriacis	1	*Neobubastes*	1
Calotemognatha	1	*Neobuprestis*	1
Castiarina	116	*Neocuris*	19
Chalcophorotaenia	4	*Neospades*	4
Chrysobothris	5	*Notobubastes*	1
Cyrioides	1	*Notographus*	4
Diadoxus	3	*Pachycisseis*	1
Dinocephalia	4	*Paracephala*	3
Diphucrania	27	*Prospheres*	1
Ethonion	5	*Pseudanilara*	6
Euryspilus	2	*Selagis*	8
Germarica	3	*Stanwatkinsius*	5
Hypocisseis	3	*Stigmodera*	3
Julodimorpha	1	*Synechocera*	2
Meliboeithon	3	*Temognatha*	23

Australia but also including New South Wales (*Burnsiellus trisulcatus*, *Stigmodera sanguinosa*) or Victoria (*Burnsiellus marmoratus*, *Neobuprestis perroni*). Several species are known mainly from Western Australia and South Australia but with records also from the Northern Territory (e.g. *Castiarina lepida*, *C. quadrifasciata*, *Chalcophorotaenia beltanae*), and *Melobasis monticola* provides an interesting South Australia–Victoria–Tasmania distribution record. *Nascio vetusta* and *Nascioides parryi* are recorded only from the south-east and are solitary outliers of genera and species otherwise centred on the east coast (but with *Nascio chydaea* occurring in Western Australia). Those species with wider ranges – those extending across the southern region of the continent, or with extensions to Queensland and/or Tasmania – are numerous (e.g. see Bellamy 2002; Barker 2006a; Lang 2020). They include *Agrilus assimilis*, *A. aurovittatus*, *A. australasiae*, *A. hypoleucus*, *Astraeus badeni*, *A. irregularis*, *A. jansoni*, *Astraeus navarchis*, *Bubastes barkeri*, *B. globicollis*, *B. inconstans*, *Chalcophorotaenia exilis*, *Chrysobothris mastersii*, *Chrysobothris*

subsimilis, *Diadoxus erythrurus*, *Diphucrania acuducta*, *D. duodecimmaculata*, *D. leucosticta*, *D. notulata*, *Ethonion corpulentum*, *Hypocisseis suturalis*, *Melobasis gratiosissima*, *M. nervosa*, *M. nobilitata*, *M. thoracica*, *Merimna atrata*, *Neocuris crassa*, *N. cuprilatera*, *N. guerinii*, *N. pauperata*, *Neospades chrysopygia*, *Pseudanilara pilosa*, *Selagis aurifera*, *S. viridicyanea*, *Stigmodera macularia*, *Temognatha fusca*, *T. heros*, *T. mitchellii*, *T. parvicollis* and *T. westwoodii*. Not surprisingly, owing to their numerical dominance, *Castiarina* is represented by numerous widely distributed species, of which *C. australasiae*, *C. erythromelas*, *C. erythroptera*, *C. flavopicta*, *C. rufipennis* and *C. thomsoni* are but a small sample.

Although the diverse *Castiarina* contains many resplendent patterned species, the related stigmoderine genus *Temognatha* has historically drawn particular attention, the large and often colourful species avidly sought. Species of *Temognatha* may emerge as early as October (*T. fusca*, *T. wimmerae*), more so in November (e.g. *T. congener*, *T. flavicollis*, *T. flavicincta*, *T. parvicollis*, *T. pubicollis*, *T. stevensii*), with the majority of species

Table 7. Examples of South Australian regional fauna groupings.

WA–SA	WA, SA–NWVic	SA–NWVic
Agrilus echidna		
Agrilus kangaroo		
Astraeus obscurus	*Astraeus major*	
Bubastes leai		*Bubastes vagans*
Bubastes subflavipennis		
	Diadoxus jungi	
Castiarina aurantia	*Castiarina aurantiaca*	*Castiarina bicolor*
Castiarina booanyia	*Castiarina carminea*	*Castiarina colorata*
Castiarina browningi	*Castiarina castelnaudi*	*Castiarina crux*
Castiarina chinnocki	*Castiarina cyanipes*	*Castiarina jekelli*
Castiarina crockerae	*Castiarina distinguenda*	*Castiarina kiatae*
Castiarina domina	*Castiarina hateleyi*	*Castiarina malleeana*
Castiarina danesi	*Castiarina marginata*	*Castiarina media*
Castiarina elderi	*Castiarina pallidipennis*	*Castiarina obscura*
Castiarina euclae	*Castiarina recta*	*Castiarina perlonga*
Castiarina flava	*Castiarina robusta*	*Castiarina triramosa*
Castiarina flindersi	*Castiarina subnotata*	*Castiarina uncata*
Castiarina hanloni	*Castiarina subtincta*	*Castiarina viridolinea*
Castiarina nullaborica		*Castiarina vittata*
Castiarina octopunctata		*Castiarina xanthopilosa*
Castiarina powelli		
Castiarina propinqua		
Castiarina puteolata		
Castiarina rediviva		
Castiarina rubriventris		
Castiarina rufolimbata		
Chalcophorotaenia castanea		
Chrysobothris amplicollis		
Diphucrania rubricata	*Diphucrania minutissima*	*Diphucrania aenigma*
	Diphucrania modesta	*Diphucrania nubeculosa*
		Diphucrania semiobscura
Melobasis angustecostata	*Melobasis andersoni*	*Melobasis beltanensis*
Melobasis aureocyanea	*Melobasis pyritosa*	*Melobasis propinqua verna*
Melobasis igniceps	*Melobasis soror*	
Melobasis occidentalis		
Melobasis rothei		
Neocuris brownii	*Neocuris dichroa*	*Neocuris fairmairei*
Neocuris sapphira	*Neocuris discoflava*	
	Neocuris pubescens	
Selagis intercribrata		
Selagis peroni		
Stigmodera cancellata		
Temognatha ducalis		
Temognatha haematica	*Temognatha flavicollis*	*Temognatha congener*
Temognatha marginalis	*Temognatha flavocincta*	*Temognatha flavomarginata*
Temognatha mnizechii	*Temognatha wimmerae*	
Temognatha stevensii		

emerging December–February and sometimes later months of autumn (Lang 2020; A Sundholm pers. obs.). Emergence times can coincide with peak flowering of host plants (e.g. *Temognatha heros* with that of *Eucalyptus leptophylla*). There is no seasonal 'constant' period of emergence however, for annual emergence events for individual species and populations can be skewed in time or with lengthy years between (often massed) emergence episodes (*Temognatha heros*, *T. ducalis*, *T. parvicollis*, *T. stevensii*) (Lang 2020). Temperature and rainfall events influence emergence patterns but larval host physiology may also be a contributing factor (influencing maturation). Such apparent seasonal irregularity appears to be the norm elsewhere for numerous Australian *Temognatha*. A number of species are now uncommon (e.g. *Temognatha flavocincta*, *T. fortnumii*) or poorly known in the state (*T. ducalis*, *T. sanguiniventris*), but even widely distributed species may be rarely encountered (e.g. *T. flavicollis*).

The majority of South Australia's *Temognatha* species, as adults, are associated with large flowering Myrtaceae (e.g. *Eucalyptus brachycalyx*, *E. camaldulensis*, *E. dumosa*, *E. leptophylla*, *Melaleuca lanceolata*) (Bellamy *et al.* 2013; Lang 2020). Several are also recorded from Proteaceae (*Temognatha flavicollis*, *T. flavocincta*, *T. pubicollis*), Casuarinaceae (*T. flavicollis*, *T. fortnumii*, *T. mitchellii*) but with occasional records from Fabaceae, Pittosporaceae, Santalaceae and Xanthorrhoeaceae.

SOUTH-WEST WESTERN AUSTRALIA

Although many of the buprestid species of south-west Western Australia are distributed more extensively, this area is a major bioregion of buprestid endemism and diversity (see Table 8) (Figs 565–576); this holds true for other invertebrate groups, vertebrates and plants as well. The region for our purpose is roughly bounded by a line drawn from Exmouth north of Geraldton, then south-east via Meekathura, Wiluna, Hyden and Lake King to Hopetoun and Esperance on the southern coast, thus encapsulating all lands to the south and south-west of that line. This generally equates with the boundaries of the Temperate Zone South-Western Vegetation Province, this bordered by the South-Western Interzone and the Eremaean Vegetation Province (discussed in Beard 1990). Ours is an arbitrary determination capturing a diversity of coastal and hinterland landforms and their associated native vegetation communities, but also great expanses of cleared 'Wheatbelt' land given to agriculture. Within its boundaries are located a number of conservation reserves, prominent among these being Kennedy Range, Fitzgerald River, Stirling Range, D'Entrecasteaux, and Leeuwin-Naturaliste National Parks, and Toolonga, Karroun Hill and Lake Magenta Nature Reserves, and also including numerous forestry reserves (e.g. Dwellingup, Harris River, Muja, Noggerup, Wilga, Millbrook State Forests).

The vegetation of the South-Western Vegetation Province varies from tall eucalypt-dominated forests in the far south-west, grading to mosaics of heath, woodland and shrublands in response to changes in soil type, reduction in rainfall and general climatic patterns. Vegetation communities include forests of jarrah–marri (*Eucalyptus marginata*, *E. calophylla*) and karri (*E. diversicolor*) in the relatively humid and high rainfall zones of the extreme south-west, low *Banksia* woodland, *Melaleuca*-forested swamps, seasonally wet heath–swampland, coastal dune woodland, scrub heath associated with coastal sandplains, mixed 'mallee' woodland, *Melaleuca–Hakea* thickets (e.g. Moresby Range north of Geraldton) and isolated estuary associated *Avicennia marina*–rush communities. The 'Wheatbelt Region' comprises vegetation sequences of scrub–heath, woodland thickets, and shrub and saltflat-adapted plant complexes; these are largely in response to changes in soil type. The 'Mallee Region', to the south-east and adjacent the Wheatbelt, is characterised by shrub–eucalypt, so termed 'mallee', formations of multi-stemmed and single-stemmed 'marlock' *Eucalyptus* species (e.g. *E. cylindriflora*, *E. foecunda*, *E. oleosa*, *E. uncinata*; *E. falcata*, *E. flocktoniae*) commonly with an understorey dominated by one or more species of *Melaleuca*, but with examples of eucalypt woodland, scrub heath ('Kwongan') and Casuarinaceae thickets. The associated semi-desert South-Western Interzone, situated inland of the Wheatbelt and Mallee regions, includes the often cited jewel beetle collection localities of Southern Cross, Norseman, Balladonia and Zanthus. Vegetation here is predominantly dry eucalypt woodland, but with sandplain associated open heaths, eucalypt–shrubland, lake-associated 'saltbush' (*Atriplex*) communities, and 'spinifex' (*Troidia*) woodland.

The South-Western Region includes Matthews and Bouchard's (2008) 'Swan' and 'Murchison' centres of endemism (based on tenebrionid genera) and Levey's (2012) 'Southwest subregion', which identified a zone of pronounced *Melobasis* endemism. In addition, given the region's proximity to Perth it has been a ready focus for field investigation of numerous plant and animal groups.

Early collectors to the region included Horace Brown, who was resident in Perth during the 1920s and '30s, Herman Goerling and Herbert Carter. Goerling first emigrated to Victoria from Germany, settling later at 'Marloo Station' at Wurarga (the site of many new species) but then moving to Pinjarra. He exported native animals for sale, but collected beetles as a hobby, his collection eventually being donated to the Commonwealth Scientific and Industrial Research Organisation (CSIRO). Many specimens are also held by the Humbolt Museum, Germany (Barker 2006a). Carter visited Western Australia on several occasions (visited 1907, 1913 and 1926). He relates collecting from 'tea-tree' (presumably *Leptospermum*) along the river banks of South Perth, there encountering *Castiarina anchoralis*, *C. octospilota*, *C. pallidiventris* and others, then later to Kelmscott collecting *Cyrioides vittigera* from *Banksia*. But prominent among the region's resident field entomologists were Keith and Edith Carnaby (after whom *Astraeus carnabyi* and *Neocuris carnabyae* are respectively named), though they also travelled much more widely throughout Australia in search of jewel beetles. Keith (1910–94) was born in Perth but later farmed at Lake Grace, then later moved to Wilga. They constructed a museum to house their vast collection, with a second museum established at Boyup Brook. In 1987 Keith privately published the *Jewel Beetles of Western Australia*, providing an overview of the fauna encountered during their collecting trips. The publication relates the fluctuation in species from one year to another – temporal episodes of vast abundance and diversity, then followed by scarcity or absence. Written to engage the interest of school children and those with little knowledge of the family its pages give insight into the wealth of the fauna, the Carnabys' dedication to the investigation of Australia's jewel beetle diversity, and the spectacular fauna to found on flowering 'mallee' trees, but also the loss of jewel beetle habitat as areas of native vegetation fell to farming.

Polycestinae

In Western Australia the speciose genus *Astraeus* is represented by two subgenera (*Astraeus*, *Depollus*) and approximately 32 described species (*Astraeus* ~23, *Depollus* 9). Almost all are restricted to Western Australia, several of which are recorded from limited geographical areas (e.g. *Astraeus goldingi*, *A. tamminensis*, *A. vittatus*). Known adult food plants are mainly members of Casuarinaceae and Cupressaceae, but there are several divergences from this theme, notably associations with Proteaceae;

examples are *Astraeus fraterculus* (*Banksia* – in addition to *Daviesia* [Fabaceae]), *A. goldingi* (*Grevillea*, *Hakea* – in addition to *Allocasuarina* [Casuarinaceae]) and *A. meyricki* (*Dryandra*). Various species are to be found during spring and summer, sometimes as early as September (*Astraeus goerlingi*, *A. robustus*, *A. aberrans*), commonly in November and December (*A. irregularis*, *A. minutus*, *A. oberthuri*), but with numerous records as late as January and February (*A. macmillani*, *A. oberthuri*, *A. polli*, *A. tamminensis*).

Xyroscelis crocata has been recorded extensively from the south-western region, and is restricted to Zamiaceae. Individuals, once the eyesight has become attuned to their form, can be found basking on the upper surface of cycad foliage, but they may readily drop to the ground when disturbed. Methods of collection are usually by careful approach netting or by beating foliage.

A single species of *Helferella* (*H. abbreviata*) has been described from Western Australia, and was first described as a species of *Germarica* by Carter (1926); the two genera being somewhat similar in general appearance. The minute and elongate form suggests that the larvae might utilise vine stems, narrow twigs or rush stems as hosts. Species from the east coast have been collected in woodland and lowland rainforest by beating foliage.

Chrysochroinae

Members are generally large, robust metallic, often greenish bronze beetles and, though of considerable size, can be very difficult to see in the trees they inhabit. All are herbivorous or do not feed as adults. Adults are, in general, active from summer to early autumn – at the end of the season the remains of individuals occasionally are encountered below trees or on strand lines. Several species of *Chalcophorotaenia* occur within (*C. castanea*, *C. martinii*, *C. violacea*), or near to (*C. sphinx*), the south-western region. *Chalcophorotaenia castanea*, originally collected by Horace Brown and later described by Carter (1916), has been recorded on *Melaleuca* (Myrtaceae) foliage. Two species of the widespread *Lampetis* (*L. corruscans*, *L. fastuosa*) are recorded from the region but these are considered 'one-time' introductions or mislabelled specimens (Bellamy 2002). Several species of *Pseudotaenia* occur in Western Australia (Bellamy 2006). *Pseudotaenia gigas* is hosted by *Acacia acuminata*, commonly called 'jam tree'; this name referring to the jam-like sap that exudes from the tunnelling of larval cossid moths, the cerambycid beetle *Rhytiphora saundersi*, and that of

Table 8.   Examples of Buprestidae (given alphabetically) from south-western Western Australia.

Agrilus assimilis assimilis (Perth), *Agrilus aurovittatus* (Moore River National Park), *Agrilus australasiae* (Tammin), *Agrilus brevis* (Spencers Brook), *Agrilus cockatoo* (Wurarga), *Agrilus echidna* (Lake Grace, Lake Monger, Wurarga), *Agrilus hypoleucus* (Spencers Brook, Swan River), *Agrilus kangaroo* (Bunbury, Myalup, Perth), *Agrilus koala* (Bunbury, Lake Grace, Yellowdine), *Agrilus wallaby* (Cue), *Agrilus wallaroo* (Wurarga)

Anilara aeraria (Geraldton, Mullewa), *Anilara subimpressa* (Tammin)

Anthaxoschema carteri (Swan River)

Astraeus aberrans (Carnamah, Paynes Find, Quairading), *Astraeus acaciae* (Wooramel), *Astraeus badeni* (Marvel Loch, Southern Cross), *Astraeus bakeri* (Dryandra), *Astraeus carnabyi* (Lake Grace, 14 km north of Needilup), *Astraeus carteri* (Lake Grace, Lake King, Southern Cross), *Astraeus dedariensis* (Borden, Dedari), *Astraeus fraseriensis* (Dryandra, Lake Grace, South Tammin Flora Reserve), *Astraeus georlingi* (106 km south of Paynes Find, Wurarga), *Astraeus goldingi* (44 km north of Galena Bridge), *Astreaus irregularis* (Gosnells, South Tammin Flora Reserve), *Astraeus lineatus* (13 km north of Goomalling, Paynes Find, South Tammin Flora Reserve), *Astraeus macmillani* (Dryandra), *Astraeus minutus* (South Tammin Flora Reserve *Astraeus multinotatus* (vic. Kundip, Leonora, Stirling Ranges), *Astraeus meyricki* (18 km south-west of Three Springs), *Astraeus obscurus* (Fraser Range, 24 km north-east Lake Grace, Needilup), *Astraeus polli* (Dedari, South Tammin Flora Reserve, Wongan Hills), *Astraeus powelli* (Quairading, Tammin, 32 km east of Yellowdine), *Astraeus robustus* (Paynes Find), *Astraeus tamminensis* (South Tammin Flora Reserve)

Bubastes bostrychoides (Albany, Esperance Bay, Geraldton), *Bubastes carnarvonensis* (Carnarvon, 30 km north of Gascoyne Junction, 43 km west of Meekatharra), *Bubastes formosa* (Cue, 40 km west of Meekatharra, 43 km east of Perenjori), *Bubastes germari* (61 km east of Perenjori, Southern Cross, 60 km north-east of Wubin), *Bubastes kirbyi* (Beverley, Chapman River, Lake Grace), *Bubastes macmillani* (Albany, Dongara, 12 km west of Enneabba), *Bubastes subflavipennis* (Ankertell, 2 km east of Buntine, Wurarga), *Bubastes subnigricollis* (70 km south-west of Paynes Find, Wurarga, 55 km west of Yalgoo), *Bubastes vanrooni* (43 km east of Perenjori, Warrumbon, Wurarga), *Bubastes viridiaurea* (16 km north of Cue)

Burnsiellus marmorata (Hopetoun, Lake Hurlstone, Lake King), *Burnsiellus trisulcata* (vic. Lake Monger, Paynes Find Road, 44 km east of Perenjori)

Calotemognatha varicollis (Wanneroo), *Calotemognatha yarelli* (Hyden, Swan River)

Castiarina adusta (Lake Grace, 80 km east of Hyden), *Castiarina amabilis* (Albany, Perth, Stirling Range), *Castiarina antia* (Quairading), *Castiarina aurantia* (Mount Magnet, Pindar, Yalgoo), *Castiarina aureola* (Lake King, Wanneroo, Yunderup), *Castiarina azurea* (Eneabba, Mullewa, Southern Cross), *Castiarina bakeri* (Wongan Hills, Wubin, Wurarga), *Castiarina bazilisca* (Moores River, Northampton, Wialki), *Castiarina boldensis* (Wembley), *Castiarina brownii* (Ankertell, Cue, Murchison dist.), *Castiarina carnabyi* (Stirling Ranges), *Castiarina chamelauci* (Geraldton dist., Wanneroo, Yanchep), *Castiarina charientessa* (Eneabba, Green Head, Lancelin), *Castiarina chlorata* (Dongara, Geraldton, Three Springs), *Castiarina coolsi* (Burracoppin), *Castiarina crocicolor* (Bindoon, Gosnells, Swan River), *Castiarina cupreoflava* (Fitzgerald River National Park), *Castiarina darkinensis* (Little Darkin Swamp), *Castiarina decemguttata* (Bejoording, Kalbarri, Swan River), *Castiarina discolorata* (Albany dist.), *Castiarina desideria* (Ankertell, Cue, Mount Magnet), *Castiarina distantia* (4 km west of Zanthus), *Castiarina diversa* (Norseman, Wubin, Zanthus), *Castiarina domina* (Cue, Coral Bay, Geraldton), *Castiarina eneabba* (vic. Hyden, Eneabba, Murchison dist.), *Castiarina enigma* (Lake Grace, Ravensthorpe, Regan's Ford), *Castiarina eremita* (Kalbarri National Park), *Castiarina ferruginea* (Eneabba, Wialki), *Castiarina flaviceps* (Dedari, Geraldton, Yellowdine), *Castiarina gravis* (Bejoording, Lake Grace, Swan River), *Castiarina incognita* (Lake King, Yellowdine, Southern Cross), *Castiarina laena* (Cockburn Sound, Lancelin, Wembley Park), *Castiarina lepida* (Cue), *Castiarina macmillani* (Wubin, Yellowdine), *Castiarina mansueta* (Lake King, Pindar, Wanneroo), *Castiarina markgoldingi* (65 km north of Galena), *Castiarina michaelpowelli* (Baladjie Lakes,Dedari), *Castiarina nonmya* (Beverley, Swan River), *Castiarina octopunctata* (Eneabba, Kalbarri National Park, Merredin), *Castiarina phaeopus* (Bunbury, Gosnells), *Castiarina picta* (Albany dist., Paynes Find, South Tammin), *Castiarina radians* (Moondon, Paynes Find, Wurarga), *Castiarina recta* (Yellowdine), *Castiarina rufa* (Mullewa, Murchison River, Pindar), *Castiarina septemspilota* (Bullabulling, Mullewa, Southern Cross), *Castiarina serratipennis* (Cue, Shark Bay, Wurarga), *Castiarina sexnotata* (Cunderdin, Lake Grace), *Castiarina turbulenta* (Perth); *Castiarina ustulata* (Southern Cross, Yellowdine)

Chalcophorotaenia australasiae (30 km north-west of Nanutarra), *Chalcophorotaenia castanea* (Cue), *Chalcophorotaenia martinii* (Beverley), *Chalcophorotaenia sphinx* (10 km north of Kookynie, 13 km south of Leonora, Southern Cross), *Chalcophorotaenia violacea* (Cue)

Chrysobothris amplicollis (Swan River), *Chrysobothris australasiae* (Swan River), *Chrysobothris petersoni* (Wurarga)

Cyphogastra farinosa (Perth)

Cyrioides elateroides (Swan River), *Cyrioides vittigera* (Albany, vic. Cape Naturaliste, Wilga)

Diadoxus erythurus (Cranbrook, Perth, 16 km west of Wurarga), *Diadoxus regius* (Kellerberrin, Norseman, Perth)

Dinocephalia browni (Merridin), *Dinocephalia leucogaster* (Moores River), *Dinocephalia thoracica* (Beverley, Geraldton, Tammin)

Diphucrania aberrans (Bunbury, Pemberton, Woodridge), *Diphucrania augustgoerlingi* (Dedari, 9 km south of Pingelly, Wurarga), *Diphucrania bilyi* (60 km south of Norseman), *Diphucrania browni* (Dedari), *Diphucrania carteri* (Yilgarn), *Diphucrania chlorata* (Dandalup), *Diphucrania corpulenta* (Pindar, Wurarga, Yellowdine), *Diphucrania cyanea* (Eneabba, 29 km east of Geraldton, Wialki), *Diphucrania cyaneopyga* (Lake Austin), *Diphucrania elliptica* (Cue), *Diphucrania macmillani* (Wanneroo), *Diphucrania patricia* (Bunbury), *Diphucrania rubricata* (~47 km north of Galena Bridge), *Diphucrania rubriceps* (Lake Grace, Perth, 9 km south of Pingelly, Zanthus), *Diphucrania speciosa* (Bayswater, Cannington, Midland Junction), *Diphucrania stigmata* (Lake Grace), *Diphucrania viridipurpurea* (Geraldton)

Ethonion breve (Perth), *Ethonion roei* (Albany)

Euryspilus viridis (Swan River)

Helferella abbreviata (?)

Hypocisseis suturalis (Swan River)

Julodimprha saundersii (Enneabba, 12 km east of Greenhead, Mount Peron)

Lampetis corruscans (Dowerin), *Lampetis fastuosa* (Kellerberrin)

Meliboeithon confusum (Fremantle, Swan River, Yanchep), *Meliboeithon cylindricolle* (Geraldton), *Meliboeithon vitticeps* (Fitzgerald River National Park, South Perth, Watening)

Melobasis acutula, *Melobasis angusta* (Wurarga), *Melobasis barkeri* (Burracoppin, Lake Grace, Southern Cross), *M. brevis* (Wurarga), *Melobasis breviserrata* (Dedari, Mukinbudin, Wurarga) *Melobasis brittoni* (Ankertell, Wurarga, 62 km east of Yalgoo), *Melobasis costifera* (Albany, Freemantle, Lake Grace), *Melobasis costipennis* (Busselton, Cape Arid National Park, Stirling Range), *Melobasis cupreovittata westralica* (Perth, Tammin, vic. Hyden), *Melobasis dissimilis* (Dongara, Geraldton, Yallingup), *Melobasis duplexicolor* (Dongara, 20 km north of Geraldton, Mundaring), *Melobasis flexa* (Ankertell, Carnarvon, Southern Cross), *Melobasis fortipunctata* (Dedari, Southern Cross, Wurarga), *Melobasis goerlingi* (34 km north of Galena, Hines Hill, Wurarga), *Melobasis interstitialis* (Chapman River, Fitzgerald National Park, Geraldton), *Melobasis janae* (Kellerberrin, vic. Paynes Find, Wurarga), *Melobasis latecostata* (Busselton, Geraldton, Tammin), *Melobasis lathami* (Cape Naturaliste, Stirling Range, Swan River), *Melobasis marlooensis* (Wurarga), *Melobasis meyricki* (Coral Bay, Geraldton, Perth), *Melobasis planithorax* (McDermid Rock), *Melobasis powelli* (vic. Paynes Find), *Melobasis rectipilosa* (vic. Hyden, Moora, Stirling Range), *Melobasis regalis regalis* (Eneabba, Lake Grace, Wurarga), *Melobasis regalis carnabyorum* (vic. Coral Bay), *Melobasis rubromarginata* (Bunbury, Fitzgerald River, Geraldton), *Melobasis robusta* (Cunderdin, Kellerberrin, Southern Cross), *Melobasis similis* (Buntine Nature Reserve, Merriden, Wurarga), *Melobasis simulata* (Katanning, Narrogin, Wembley), *Melobasis soror basicostata* (Dongara, Mon Repos, Perth)

Nascio chydaea (Boddington, Bridgetown, Stirling Range)

Neobubastes aureocincta (Yellowdine), *Neobubastes flavovittata* (Cunderdin, Kellerberrin), *Neobubastes nickerli* (Dedari, 10 km north of Pindar), *Neobubastes obscura* (2 km north of Bullabulling, Lake Campion, 43 km east of Perenjori)

Neocuris anthaxoides (Perth), *Neocuris brownii* (Cue), *Neocuris carnabyae* (Coral Bay, 120 km south of Onslow), *Neocuris discoflava* (Mandurah), *Neocuris duboulayi* (Three Springs), *Neocuris ignicollis* (Wurarga), *Neocuris thoracica* (King George Sound), *Neocuris viridiaurea* (King George Sound), *Neocuris viridimicans* (Geraldton, King George Sound)

Notobubastes occidentalis (Cue)

Pseudanilara dubia (Beverley)

Pseudotaenia gigas (Paynes Find, 6 km north of Yellowdine, Swan River, Wurarga)

Selagis baumi (Wurarga), *Selagis carteri* (Wurarga), *Selagis chloriantha* (King George Sound), *Selagis despecta* (Champion Bay), *Selagis discoidalis* (Yilgarn), *Selagis intercribrata* (Swan River, Toodyay dist.), *Selagis venusta* (Wurarga), *Selagis yalgoensis* (Yalgoo)

Stanwatkinsius cariniceps (48 km east of Geraldton, Paynes Find), *Stanwatkinsius cinctus* (13 km north of Galena, Woodridge, Marloo Stn./ Wurarga), *Stanwatkinsius constrictus* (vic. Armidale, Stirling Ranges, Wilga), *Stanwatkinsius crassus* (Lake Grace), *Stanwatkinsius demarzi* (Woodridge), *Stanwatkinsius grevilleae* (13 km north of Galena, Tammin, Yellowdine), *Stanwatkinsius macmillani* (Watning), *Stanwatkinsius powelli* (Cue, 74 km south-east of Yalgoo), *Stanwatkinsius perplexus* (Kalbarri, Lake Grace, Quairading, South Tammin Flora Reserve), *Stanwatkinsius speciosus* (25 km north of Eneabba, 17 km south of Northampton), *Stanwatkinsius subcarinifrons* (3 km north-east of Gosnells, 58 km west of Tammin, Woodridge), *Stanwatkinsius viridimarginalis* (Dryandra State Forest, Swan River, 34 km east of Yellowdine)

Stigmodera cancellata (Albany dist.), *Stigmodera gratiosa* (Lake Grace, Stirling Range), *Stigmodera roei* (Arrowsmith River, Lake Bryde, Perth), *Stigmodera sanguinosa* (vic. of Ravensthorpe, Swan River)

Synechocera elongata (Canning Vale, Fremantle, Swan River), *Synechocera parvipennis* (Eneabba), *Synechocera setosa* (Geraldton, Pemberton, Swan River)

Temognatha bonvouloirii (Kalbarri, Marvel Loch), *Temognatha brevifasciata* (Swan River), *Temognatha bruckii* (between Newdegate and Lake King, 30 mi. (48 km) south of Norseman, vic. Yellowdine), *Temognatha chalcodera* (beween Newdegate and Lake King, Marvel Loch, vic. Yellowdine), *Temognatha chevrolatii* (vic. Hyden, Lake Grace, Southern Cross), *Temognatha conspicillata* (vic. Lake Grace, Swan River), *Temognatha coronata* (Yellowdine), *Temognatha ducalis* (Zanthus), *Temognatha duponti* (between Newdegate and Lake King, Lake Bryde, Ravensthorpe), *Temognatha gigas* (Southern Cross), *Temognatha gloriosa* (Murchison dist.), *Temognatha gordonburnsi* (69 km north of Galena), *Temognatha haematica* (Swan River), *Temognatha heros* (vic. Hyden, Ravensthorpe), *Temognatha imperialis* (Cue, Norseman, vic. Southern Cross), *Temognatha lessonii* (Stirling Ranges, Swan River), *Temognatha martinii* (Southern Cross-Ravensthorpe dist.), *Temognatha ducalis* (vic. Hyden, Marvel Loch, Zanthus), *Temognatha mitchellii* (Swan River), *Temognatha mnizechii* (vic. Hyden, Lake King, Norseman), *Temognatha murrayi* (41 mi. [66 km] east of Norseman, Marvel Loch, Swan River), *Temognatha nigrofasciata* (Yilgarn), *Temognatha obscuripennis* (Stirling Ranges), *Temognatha oleata* (vic. Hyden, near York), *Temognatha parvicollis* (vic. Hyden), *Temognatha pictipes* (41 mi. [66 km] east of Lake King, 50 mi. [80 km] south of Norseman, near York), *Temognatha pubicollis* (Swan River), *Temognatha punctatostriata* (Swan River), *Temognatha rectipennis* (vic. Hyden, Norseman), *Temognatha reichei* (Jandakot, Stirling Ranges, Swan River), *Temognatha secularis* (Perth, Swan River), *Temognatha stevensii* (Kundip, Stirling Range, Zanthus), *Temognatha westwoodii* (Yellowdine), *Temognatha wimmerae* (vic. Hyden, Norseman, Southern Cross dist.).

Xyroscelis crocata (Kings Park, Swan River, Warren River, Wanneroo, Melville, Southern Cross)

Pseudotaenia gigas itself. Mark Hanlon (in Bellamy 2006) describes adults during mid-morning landing on the crown of *Acacia acuminata*, with a female observed walking down the branches onto the main trunk, the beetle's abdomen extended slightly, and then rubbing the abdomen on the trunk as if attempting to locate a suitable site to oviposit. Although *Acacia acuminata* is a larval host plant, adults have been collected feeding from eucalypt leaves during the months of December and January.

Buprestinae

The region is especially rich in species of *Melobasis*, the majority being endemic, with most of the species occurring in the rainfall zone that receives 300–800 mm annually (Levey 2012). Although many species are widely recorded (Levey 2012, 2018; Table 8) there are several with very limited known localities (e.g. *Melobasis marlooensis*, *M. planithorax*). Adults are generally encountered from August to November, but a number of species are also present in summer (*Melobasis latecostata*, *M. soror basicostata*). There are frequently peaks of emergence within more seasonally extensive presence, as with the common *Melobasis lathami* that has been collected from November to March but with most records in December and January (Levey 2018). Most known adult host plant associations are with *Acacia* (Fabaceae) leaves, also flowering plants (*Melobasis barkeri*, *M. interstitialis*); however, additional adult host plant records include the fabaceous genera *Chlorizana* (*Melobasis rectipilosa*), *Gastrolobium* (*M. uniformis*), *Jacksonia* and *Viminaria* (*M. rubromarginata*), and *Mirbelia* (*M. r. regalis*, *M. simulata*, *M. uniformis*), *Leucopogon* – Eriaceae (*M. uniformis*). Few larval hosts are known.

Beetles of the cylindrical genus *Bubastes* are particularly well represented. Adults may emerge as early as October (*Bubastes subflavipennis*) but are generally present in December to February (*B. germari*, *B. vanrooni*, *B. viridiaurea*), being mostly associated with *Eucalyptus* and *Acacia* leaves. However, individuals have been collected in blue and green traps (e.g. *Bubastes macmillani*, *B. viridiaurea*) (Bílý and Hanlon 2020). The related *Neobubastes obscura* has been reared and cut from the trunks and timber billets of *Melaleuca lateriflora* and *M. uncinata* (Bílý and Powell 2017); the genus *Melaleuca* (*M. viridiflora*) is also used as a larval host by species of *Bubastes* (Bellamy *et al.* 2013). Adults of *Neobubastes obscura* have been collected in October, November and February. *Neobubastes aureocincta* has been collected on the foliage of *Acacia* and *Eremophila* and also on the flowers of *Grevillea*. Bellamy and Peterson (2000a) cite adult *Neobubastes aureocincta* being collected during September–November, with adults of *N. flavovittata* and *N. nickerli* being active in late summer.

Of the four species of *Nascio* recorded from Australia, only *Nascio chydaea* occurs in, and is restricted to, southwestern Western Australia. Adults have been encountered in the months September and November–February. They are coppery bronze with two reddish fasciae placed midway and in the apical third of the elytra, these reaching similarly coloured elytral margins that extend from near the humeral angles almost to the apex. Recent specimens have been collected 11 km south of Boddington in eucalypt woodland, in the Stirling Ranges (on an undetermined Asteraceae flower) and at Bridgetown (here breeding in *Eucalyptus wandoo*). Otherwise, the species appears to remain poorly known. The designated types (a male and female lodged in the Australian Museum, Sydney) of *Nascio chydaea* (Fig. 81) were collected at 'Salt River', which Peterson (2018) considers refers to 'Salt River (= Pallinup River) at Mongup homestead (34°11′S, 118°19′E) on Peenebup Creek, a tributary of Salt River on its upper reaches'. The original Western Australian specimens were believed to have been collected by George Masters in 1869 (Peterson 2018). The species was described in 1886 by Sidney Olliff, then Assistant Zoologist (Entomology) at the Australian Museum, but he mistakenly included two distinct species (from Western Australiain and eastern Australia) in the type series. The second species was described by Williams and Watkins in 1985 as *Nascioides olliffi* (Fig. 92); this is currently known only from New South Wales and south-eastern Queensland, and associated with Casuarinaceae.

Julodimorpha includes two species: *J. saundersii* from Western Australia and *J. bakewellii* from mainland south-eastern Australia. Adults have been recorded in spring, and have been observed ovipositing in *Calothamnus* sp. (Myrtaceae). Hawkeswood and Knowles (1985) record the dismembered remains of *Julodimorpha saundersii* in coastal *Banksia* heathland at Mount Peron, having apparently been preyed upon by birds. Carnaby (1987) provides an interesting photograph of massed numbers of male *Julodimorpha saundersii* (as *J. bakewelll* as it was then understood) attempting to mate with a beer bottle on the road verge near Eneabba. He notes observing as many as 26 individuals attracted to a single bottle and that beetles would usually stay until they died, or fell prey to crows and foxes.

The genus *Cyrioides* comprises six described species, four from eastern Australia and two (*C. elateroides*, *C. vittigera*) known only from south-western Western Australia; this distribution mirroring the faunal disjunction in *Nascio*, *Julodimorpha* and *Xyroscelis* (Polycestinae). Bellamy (2002) gives the type locality of *Cyrioides elateroides* as 'Swan River', but the probable holotype of *C. vittigera* is known only as 'New Holland', a common labelling generality with specimens collected early in Australia's entomological history. *Cyrioides vittigera* has been collected in open woodland near Bunker Bay (vic. Cape Naturaliste) in association with *Astraeus fraterculus* on the foliage of *Banksia attenuata*. Carnaby (1987) mentions that the species inhabits forest areas near Wilga where he found it on *Banksia grandis* in January, and at Albany on a white flowering *Hakea* in the same month.

Diadoxus erythrurus and *D. regius* (the larger of the two) are distributed widely in the south-west (Peterson and Hawkeswood 1980). Adult *Diadoxus* are associated with *Callitris* (*C. endlicheri*, *C. huegelii*, *C. verrucosa*) (Cupressaceae), but also with introduced *Cupressus* (Peterson and Hawkeswood 1980; Hawkeswood and Turner 1997), their colour blending with that of the host foliage. Adults of both species may be found in the region during late spring and summer.

Bellamy (2002) cites seven species of *Chrysobothris* occurring in Western Australia. Of these *Chrysobothris amplicollis*, *C. mastersii*, *C. saundersii* and *C. subsimilis* exhibit distributions also with eastern states, with *C. australasiae*, *C. macleayi* and *C. petersoni* having more circumscribed Western Australian distributions. Hawkeswood (1995) described *Chrysobothris petersoni* from a unique female specimen collected by Hermann Goerling at the often cited locality 'Marloo Station, Wurarga'. As with many other Australian *Chrysobothris* its larval habits are unknown.

The *Selagis* fauna includes *S. baumi*, *S. carteri*, *S. chloriantha*, *S. despecta*, *S. discoidalis*, *S. intercribrata*, *S. spencei*, *S. venusta* and *S. yalgoensis*, and though species are widely known to be florivorous few adult host records have been published (Bellamy *et al.* 2013). *Selagis intercribrata*, also recorded from South Australia (Lang 2020), has been reared from the stems of Casuarinaceae (though the species is uncertain [discussed in Hawkeswood 1988]).

At least 12 species of *Neocuris* are recorded from Western Australia, but only a few appear to be endemic (e.g. *N. brownii*, *N. carnabyae*, *N. duboulayi*, *N. ignicollis*). All species are probable nectar feeders. As previously noted *Neocuris discoflava* has been recorded from flowering *Nuytsia floribunda* (Loranthaceae), and there is mention by Carnaby (1987) of the species being found on 'Rottnest Island daisy', *Trachymene coerulea* (Apiaceae), at Mandurah, but otherwise few records have been published.

Of the about 30 species of the diminutive sized *Anilara* listed from Australia by Bellamy (2002) only *A. adelaidae*, *A. aeraria* and *A. subimpressa* are cited by him from Western Australia. H. J. Carter's 1926 'revision' of the genus provides little information regarding host and habitat associations, or distribution details. Adult *Anilara* can be found in late spring and summer associated usually with the foliage of freshly fallen eucalypt branches, in which females oviposit their eggs. Individuals can be collected by beating the boughs and dying foliage, but dry and brittle leaves are usually devoid of beetles; the response of beetles to leaves usually occurs several days after bough fall and subsequent attraction lasts for about a week. Despite the ease of collecting adults there is surprisingly little published on host associations, habitat and distribution. This holds for the Australian *Anilara* fauna in general (Bellamy *et al.* 2013). The genera *Notographus* and *Pseudanilara* are represented at least by two (*N. uniformis*, *N. yorkensis*) and three (*P. dubia*, *P. occidentalis*, *P. occidentis*) species respectively (Bellamy 2002) and in similar fashion to *Anilara* can be collected by beating foliage and tree boughs.

Stigmoderinae

But it is the buprestid fauna associated with the region's spectacular wealth of wildflowers, flowering 'mallee' and myrtaceous species in general that has drawn the attention of entomologists and collectors alike, and it is the Stigmoderinae that draw much of the buprestid attention. Four genera are represented in the south-west: *Calotemognatha*, *Castiarina*, *Stigmodera* and *Temognatha*. *Calotemognatha* is represented by *C. laevicollis*, *C. varicollis*, and *C. yarelli*, although the genus extends to Victoria and New South Wales (Peterson 1991). Adults have similar flower-frequenting habits to regional *Temognatha*. Host plants for adult *Calotemognatha yarelli* include *Nuytsia floribunda*, *Eucalyptus foecunda*, *E. uncinata*, *Melaleuca acuminata* and *M. thymifolia*.

There are four species of *Stigmodera* represented, three of which (*S. cancellata*, *S. gratiosa*, *S. roei*) are restricted to Western Australia, but with a fourth (*S. sanguinosa*) also occurring in South Australia and Victoria. *Stigmodera gratiosa* is a brilliant metallic green–blue beetle whose distribution extends throughout the south-west. Adults can be encountered from late winter to summer, their flowering hosts including *Agonis*, *Eucalyptus* and *Melaleuca* (Myrtaceae) as well as *Hakea* (Proteaceae) (Bellamy *et al.* 2013). *Stigmodera roei* and *S. cancellata* are similarly coloured metallic green with elytral margins and subfascia coloured orange–red. Adults generally emerge September–November and can be found on flowering Myrtaceae (e.g. *Chamelaucium uncinatum*, *Eucalyptus* and *Leptospermum* spp.). *Stigmodera sanguinosa* shares with *S. jacquinotii* from eastern Australia conspicuously spinose elytral apices. It is a distinctively coloured species in which the elytra are a deep blood red contrasted with numerous dark-toned punctures, and occurs from late winter to about November. Adults are attracted to *Nuytsia floribunda* (Loranthaceae) as well as species of *Leptospermum* and *Melaleuca*.

The genus *Castiarina* is prolific, but only a small proportion of the known species (e.g. *C. distinguenda*, *C. dugganensis*, *C. pallidiventris*, *C. recta*) extend in their range to the eastern coastal states. Species may emerge as early as July–August (*C. acuticeps*, *C. brownii*, *C. domina*), commonly in early–mid-spring (*C. charientessa*, *C. ferruginea*, *C. macmillani*) and extending into summer (*C. markgoldingi*, *C. michaelpowelli*). Although a number can be common or widely collected (*C. anchoralis*, *C. aureola*, *C. azurea*, *C. filiformis*) there are species with more circumscribed distributions (*C. boldensis*, *C. coolsi*, *C. distantia*, *C. eremita*, *C. markgoldingi*). A number are threatened owing to land clearing (e.g. *Castiarina carnabyi*). Many new species were first collected by Horace Brown (*C. boldensis*, *C. browni*, *C. coolsi*, *C. desideria*, *C. subtincta*) during his residence in Western Australia during the early 1900s. As previously discussed there is an association of adult *Castiarina* particularly with flowering Myrtaceae, with episodic emergence events in eucalypt-dominated 'mallee' communities being occasions of renown. However, there are a number of host divergences (though often retaining an association with Myrtaceae), examples being *Castiarina amabilis* (*Nuytsia floribunda*), *C. browningi*, *C. diversa*, *C. rufa* (*Eremophila* – Scrophulariaceae), *C. laena* (*Helichrysum* – Asteraceae) and *C. markgoldingi* (*Dicrastylis* – Lamiaceae). *Castiarina chinnocki* exhibits an interesting host association that changes with respect to its distribution, with adults in Western Australia being found on *Eremophila* and those in South Australia occurring on *Myoporum* – though both genera are in the family Scrophulariaceae. Floral fragrances serve as primary agents of attraction but colour, or 'contrast', may function as a recruitment signal, as suggested by individuals being collected in coloured buckets (e.g. as the type series of *Castiarina darkinensis* being caught in a red buckets implies). A number of species are difficult to distinguish owing to their similarity of form and colour (*Castiarina adusta*, *C. azurea*, *C. charientessa*, *C. ferruginea*, *C. nonyma*, *C. phaeopus*, *C. ustulata*) (Barker 1996).

But it is, arguably, *Temognatha* that serves as an icon for Western Australia's buprestid fauna. Given their size, flight and feeding habits adult *Temognatha* are likely to be important agents of pollen transfer and out-crossing. About 40 species of the genus are recorded from Western Australia (Bellamy 2002), the majority being endemic to the state's borders, though with several extending to South Australia or Victoria (e.g. *T. duponti*, *T. flavicincta*, *T. pascoei*), News South Wales (*T. heros*),

Queensland (*T. mnizechii*, *T. parvicollis*, *T. wimmerae*) or Tasmania (*T. mitchellii*) (Bellamy 2002). Individuals can be a seasonally conspicuous element of the adult insect fauna associated with flowering 'mallee' (the larvae being associated with roots), especially so in mid–late summer (often a time of excessively hot days), though also in spring (e.g. *Temognatha flavicincta*, *T. marginalis*). Despite severe clearing of native vegetation rich areas of mallee vegetation communities survive in the south-west wheatbelt, and when seasonally in flower the many-stemmed and single-stemmed eucalypts may attract numerous species to individual trees (Sundholm and Catford 1982). The diversity of co-occurring *Temognatha* beetles can be spectacular with species such as *Temognatha chevrolatii*, *T. ducalis*, *T. heros*, *T. mnizechii*, *T. oleata*, *T. parvicollis*, *T. pictipes*, *T. rectipennis* and others encountered at individual locations and frequently in the canopies of individual trees. Keith Carnaby in *Jewel Beetles of Western Australia* (1987) relates finding in 1970, in the vicinity of Hyden, 12 species together on one small mallee tree. This event was followed at the same location by about 10 years of varied abundance and occasional years of absence, then in 1980, during January and February, massed emergences of *Temognatha* species again occurred.

Although individual species may be widespread and abundant in some seasons (e.g. *Temognatha bonvouloiri*, *T. bruckii*, *T. chevrolatii*, *T. ducalis*, *T. murrayi*, *T. pictipes*), nevertheless there are those that are generally considered rare or uncommon, or relatively localised in known distribution (*T. conspicillata*, *T. coronata*, *T. imperialis*, *T. lessonii*, *T. reichei*, *T. westwoodii*). Most recorded adult *Temognatha* flower host associations are from *Eucalyptus foecunda* and *E. uncinata* (e.g. *Temognatha bonvouloirii*, *T. chalcodera*, *T. chevrolatii*, *T. ducalis*, *T. heros*, *T. ducalis*, *T. mnizechii*, *T. murrayi*, *T. parvicollis*, *T. rectipennis*), with additional flowering eucalypt host species including *Eucalyptus cylindriflora*, *E. gracilis*, *E. leucoxylon*, *E. oleosa*, and *E. pauperiflora* (see Bellamy et al. 2013). There are several records from flowering *Melaleuca* (*Temognatha chevrolatii*, *M. lessonii*). However, as also exhibited by the related *Castiarina*, there are departures from this Myrtaceae flower association, including *Temognatha lessonii*, *T. obscuripennis*, and *T. secularis* on *Nuytsia floribunda* (Loranthaceae) and *T. gordonburnsi* on *Bursaria spinosa* (Pittosporaceae/Bursariaceae) and *Grevillea* (Proteaceae). Larvae are associated with plant roots, but published larval host records are relatively few, the difficulty

in raising pre-adult stages to maturation being understandable. Records include *Temognatha chalcodera* (*Allocasuarina acutivalvis*), *T. heros* (*Allocasuarina/ Casuarina*, *Eucalyptus foecunda*, *E. gracilis*, *E. oleosa*, *E. uncinata*, *Melaleuca uncinata*), *T. imperialis* (*E. striaticalyx*), *T. martinii* (*Allocasuarina corniculata*), *T. mnzechii* (*E. foecunda*, *E. gracilis*, *E. oleosa*, *E. uncinata*) and *T. parvicollis* (*E. foecunda*, *E. gracilis*, *E. oleosa*, *E. uncinata*) (Bellamy et al. 2013).

Agrilinae

The *Agrilus* fauna of the south-western region comprises at least 11 species (Curletti 2001; Bellamy 2002). *Agrilus cockatoo*, *A. koala*, *A. wallaby* and *A. wallaroo* are restricted to Western Australia but several have distributions that extend to South Australia (e.g. *Agrilus assimilis assimilis*, *A. echidna*, *A. kangaroo*), New South Wales and/ or Queensland (*A. aurovittatus*, *A. australasiae*, *A. brevis*), and Tasmania (*A. hypoleucus*). Adults are active in summer (e.g. *Agrilus echidna*, *A. kangaroo*, *A. koala*) and have been recorded in association with the foliage of *Acacia* (e.g. *A. australasiae*, *A. echidna*, *A. kangaroo*, *A. koala*), *Allocasuarina* and *Leptospermum* (*A. echidna*) (Curletti 2001; Bellamy et al. 2013).

Although *Synechocera*, Australia-wide, is only a small genus of only nine species, the south-western region of Western Australia has three species. *Synechocera elongata* is the most commonly encountered Western Australian species. Eastern species of this distinctively flattened genus are associated with *Gahnia* (Cyperaceae) but Robert McMillan collected adults of *S. parvipennis* in December from the leaves of 'grass trees' *Xanthorrhoea* (Xanthorrhoeaceae) (Bellamy 1987b).

In a contrast of form to adult *Synechocera*, the cylindrical *Meliboeithon vitticeps* is one of several species in the genus recorded from Western Australia, *Meliboeithon* extending across the mainland quite 'littorally' from the shores of south-western Western Australia to the estuarine rush swamps of the east coast. Unfortunately the life history of *Meliboeithon vitticeps*, and that of most other members of the genus, is poorly known. The related *Meliboeithon cylindricolle* occurs in Western Australia, South Australia and New South Wales, with adults collected in both Western Australia and New South Wales from pitfall traps (Bellamy 1988). Bellamy considered that the collection of specimens in pitfall traps suggested 'a larval biology which utilises the root crown or stem base of the host plant, with emergent adults flying very low to feeding and mating sites'. The related genus *Dinocephalia* includes *D. thoracica*, *D. browni* and *D. leucogaster* from the southwest, of which only *D. thoracica* is more widely distributed, being also recorded from Queensland, New South Wales, South Australia and Tasmania (Bellamy 1988). In New South Wales the species has been collected commonly on the foliage of *Allocasuarina distyla* (Casuarinaceae) and a single record by Bellamy from flowers of *Bursaria* (Pittosporaceae/Bursariaceae), but otherwise host records are unknown.

More than 30 described species of *Diphucrania* are recorded from Western Australia (Barker 2001, 2002, 2006c), many of which are known from the south-west (Table 8) but a number of species also extend more widely on the mainland (e.g. *D. duodecimmaculata*, *D. parva*, *D. roseocuprea*, *D. rubicunda*). Adults are active commonly during late spring–summer (e.g. *D. augustgoerlingi*, *D. corpulenta*, *D. speciosa*) but also may be encountered as early as June–July (*D. aberrans*, *D. cyanea*, *D. speciosa*) and later in autumn (*D. macmillani*, *D. rubriceps*). They can be collected by beating foliage or by careful observation. *Diphucrania aberrans* has been collected in intercept traps (Barker 2001). Species may be found perching or feeding on foliage (often that of *Acacia*, e.g. *D. augustgoerlingi*, *D. rubricata*), or sometimes in association with flowers. *Diphucrania rubriceps* is recorded from the foliage of *Acacia* and *Grevillea*, but also from eucalypt flowers, *D. corpulenta* has been collected mostly from flowering *Melaleuca*, *D. viridipurpurea* on the foliage of *Chamelaucium uncinatum* (Myrtaceae), *D. rubricata* has been collected from the foliage of *Allocasuarina campestris*, and *Diphucrania macmillani* has been collected from the foliage of *Banksia attenuata* and *Xanthorrhoea*. *Diphucrania cyanea* is known to make galls in *Acacia blakelyi*, with adults also occurring on *A. blakelyi* flowers (Bellamy *et al.* 2013).

Species of *Stanwatkinsius* are of similar form to *Diphucrania*, with several being originally ascribed to *Cisseis* (Barker and Bellamy 2001). Most Western Australian species are endemic to that state. Several occur in southwestern Western Australia and adjacent inland areas. Adults are normally associated with foliage and have been collected from September to January but with an apparent peak in October–November. Adult host plant records are mainly from the foliage of Casuarinaceae and Proteaceae; this is a separating distinction from species of *Diphucrania* which are mainly associated with *Acacia* (Fabaceae). The following host records are from Barker

and Bellamy (2001): *Stanwatkinsius careniceps* (associated with *Allocasuarina campestris*, most common on the northern edge of the 'Wheatbelt'), *S. cinctus* (*Hakea* spp.), *S. constrictus* (*Grevillea*, *Hakea*, and '*Casuarina*'), *S. crassus* (*Grevillea*), *S. grevilleae* (*Grevillea*, *Hakea*, '*Casuarina*'), *S. perplexus* (common in drier heath and associated with *Allocasuarina* spp.), *S. powelli* (*Grevillea*), *S. speciosus* (*Hakea* spp., coastal plain), and *S. subcarinifrons* (*Allocasuarina* spp.). In addition *Stanwatkinsius demarzi* has been collected in intercept traps set in *Banksia* woodland (at Woodridge) during October, November and December (H. Demarzi records in Barker and Bellamy 2001). Barker and Bellamy (2001) highlight that *Stanwatkinsius macmillani* sp. nov. was known to them only by specimens collected from Watning and Bejoording by Robert McMillan in November 1950, the two localities 'devastated by land clearance for agriculture'.

TASMANIA

The vegetation of Tasmania ranges from coastal and alpine heath (Figs 559, 560), open woodland and eucalypt forests, to iconic mesic eucalypt forests and cool temperate rainforests, and offers a unique opportunity to consider the complete known buprestid fauna of a temperate island. In doing this a dramatic difference is to be seen when comparing the fauna of Tasmania (about 6 800 000 ha in area), with that of New Zealand (26 800 000 ha). Fifty-two species of Buprestidae are recorded from Tasmania, however, New Zealand possesses but two buprestids: *Maoraxia erimita* and *Nascioides enysii*. These two genera reside within the subfamily Buprestinae but beyond that are separately placed (see Hołyński 1984, 1988; Bellamy 2002, 2003). Physical disjunction from other Australo-Antarctic Gondwanan landmasses about 80 mya, and submersion of the New Zealand landmass in the Oligocene (Mildenhall *et al.* 2014), may go some way to explain the impoverished state of New Zealand's buprestid fauna, but remains enigmatic. The larval host plants of *Nascioides enysii* are five species of *Nothofagus* (Nothofagaceae), *N. enysii* thus having a very narrow host plant dependency. This obligate host relationship lends support to the suggestion that Oligocene 'islands' may have escaped inundation (see discussion in Mildenhall *et al.* 2014). The genus *Maoraxia* is widespread within the south-western Pacific, its distribution indicating significant dispersal and speciation episodes, with the occurrence of *M. erimita* possibly representing a post Late Oligocene colonisation event.

In comparison to that of New Zealand, the Tasmanian fauna is represented by four subfamilies: Polycestinae (*Helferella* [1 sp.]); Buprestinae (*Cyrioides* [1 sp.], *Melobasis* [12 spp.], *Nascioides* [2 spp.], *Pseudanilara* [4 spp.]); Stigmoderinae (*Castiarina* [21 spp.], *Temognatha* [1 sp.]) and Agrilinae (*Agrilus* [1 sp.], *Diphucrania* [5 spp.], *Dinocephalia* [1 sp.], *Ethonion* [3 spp.], *Synechocera* [1 sp.]) (Cowie 2001; Grove and Yaxley 2004; Fearn 2016; Richards and Spencer 2021). This represents a fauna largely reflective of the mainland, but with notable absences, especially *Astraeus*, *Stigmodera* and all of Chrysochroinae. *Temognatha* is represented only by *T. mitchellii*, a common species from coastal heathlands in eastern Tasmania, and widely recorded from the mainland. *Castiarina* is well represented, though some, such *C. rufipennis*, *C. bifasciata* and *C. undulata*, may not be permanent residents (Cowie 2001; Richards and Spencer 2021). In addition to a suite of species widespread on the eastern mainland (e.g. *Castiarina australasiae*, *C. bifasciata*, *C. bremei*, *C. cruentata*, *C. erythromelas*) and species with a more limited mainland distribution (*C. insularis*, *C. leai*, *C. thomsoni*), there are several endemic species whose isolated presence suggests speciation following the opening of Bass Strait, namely *Castiarina insculpta*, *C. jubata*, *C. macquillani*, *Castiarina ocelligera*, *C. rudis*, *C. tasmaniensis* and *C. virginea* (Cowie 2001; Barker 2006a). Of these *Castiarina ocelligera* (Fig. 152), a possible cantharid mimic, and *C. tasmaniensis* (Fig. 455) are widespread in non-alpine areas, *C. virginea* (Fig. 463) is widespread but uncommon, *C. jubata* (Figs 397, 398) is uncommon (Corinna, Waratah, Meredith River, Lake Dobson), and *C. macquillani* (Mt Algonkian, Mt Doris) and *C. rudis* (Lake Dobson, Great Lake, Lake St Clair, Wineglass Bay) are rare. *Castiarina macquillani* (Fig. 410) has been collected on *Leptospermum ruprestre*, during summer in upland moors and heathland, but its life history is otherwise unknown. *Castiarina rudis* (Figs 159, 473), which Barker (1980) considered a lycid mimic, occurs in low densities in upland heathland, with adults recorded 'flying only on very hot days in late summer' (Cowie 2001). *Castiarina insculpta* (Figs 391, 392) inhabits heath and open woodland and is listed as '*endangered*' (though was first thought to be extinct) under the Tasmanian *Threatened Species Protection Act 1995* (Richards and Spencer 2021). Cowie (2001) records it first being collected in 1934, and then a second specimen in 1965 (that is in the South Australian Museum collection). However, Richards and Spencer (2021) later located additional specimens in

Tasmania's central highlands, the pre-adult stages being associated with *Ozothamnus hookeri* (Asteraceae), with adults sometimes feeding on blossoms of *Baeckea gunniana* (Myrtaceae). Flowering *Leptospermum* species are also possible adult food plants. *Castiarina jubata* inhabits open heath and montane moorland communities, and occurs on flowering *Leptospermum*. The pronotum and head of *Castiarina jubata* and *C. macquillani*, and the pronotum of *C. rudis*, are distinctly setose, their form bringing to mind members of the related Chilean genus *Lasionota* (previously referred to as *Dactylozodes*).

Nascioides is represented by the rainforest-inhabiting *N. quadrinotatus*, and the conspicuously orange–black *Nascioides parryi*. The latter is found in open forest, as it does on the mainland (South Australia, Victoria, New South Wales). *Nascioides quadrinotata* also occurs in Victoria. Its larval host is *Nothofagus cunninghamii*, this tree restricted to Tasmania and Victoria (Hnatiuk 1990). There is also a specimen of *Nascioides carissimus* labelled as from Tasmania in the Macleay Museum Collection, but without a locality (Williams 1987). This may be a wrongly labelled specimen. The four species of *Pseudanilara* (*P. cupripes*, *P. kerremansi*, *P. piliventris*, *P. purpureicollis*) have very limited recorded distributions in the south-east (e.g. Kingston) (Cowie 2001). *Pseudanilara kerremansi* is an endemic (Bellamy 2002). *Melobasis* includes the metallic-hued *M. hypocrita* and the attractively marked *M. monticola*. Both inhabit open woodland, but interestingly *M. hypocrita* is also found in cool temperate rainforest in northern New South Wales (Williams 2020a). Cowie (2001) recorded *Melobasis propinqua porteri* as a uniquely Tasmanian subspecies. There is a single species of *Synechocera* (*S. deplana*, cited as its synonym *S. tasmanica* in Cowie [2001]).

The island state has long been the lure of entomologists. The majestic panoramas of its unspoiled (and uncleared and unlogged) forests and landscapes need little by way of hyperbole. H. J. Carter visited Tasmania in January 1901. He and a companion journeyed both sides of the island, and they did this by bicycle. Along the way, in the vicinity of the 'Blue Tiers' in Tasmania's north-east, and while resting by dense forest that lined an especially steep grade, Carter was gifted with the sight of *Thylacinus cynocephalus* – the thylacine, alternatively known as the Tasmanian tiger and Tasmanian wolf. It was a fleeting view, and even then the carnivorous marsupial had become a creature of rarity. Carter wrote about it in *Gulliver in the Bush*, its date of publication in 1933 approximating the last capture of a living animal, and pre-dating the thylacine's extinction by only 3 years.

Carter returned in January 1902, this time to attend a congress in Hobart, and there collected his first Tasmanian beetles, paying particular attention to the *Leptospermum* that flowered on the slopes of Mount Wellington. His companions on this occasion were A. H. Lucas and J. M. Stephen. Among the many beetles collected were the buprestids *Castiarina ocelligera*, *C. australasiae* and *Nascioides parryi* and at Port Arthur he collected *Castiarina wilsoni*. But a later visit in 1917, this time with Arthur Lea, proved his most productive of visits, at Sheffield collecting *Nascioides quadrinotatus* from *Acacia melanoxylon*, and at Burnie in the north *Castiarina dimidiata* (though this possibly being *C. leai*), *C. flavopicta*, *C. cyanicollis* var. *viridis* (possibly misidentified) and *C. ocelligera*.

COLOUR SECTION 3: HABITAT

Fig. 523. Rainforest canopy, Wet Tropics (Qld). ©Hugh Nicholson

Fig. 524. Rainforest-eucalypt forest ecotone with tall flowering *Leptospermum* sp. West of Paluma (NQld). ©Allen M. Sundholm

Fig. 525. Tropical rainforest, Paluma (NQld). ©Allen M. Sundholm

Fig. 526. Mixed sclerophyll forest, Kroombit Tops (CQld). ©Geoff Williams

Fig. 527. Mixed woodland, Mt Moffatt, Carnarvon National Park (Qld). ©Hugh Nicholson

Fig. 528. Mt Moffatt, Carnarvon National Park (Qld). ©Hugh Nicholson

Fig. 529. *Araucaria cunninghamii* (Araucariaceae) (foreground), larval host for *Araucariana queenslandica*, vine scrub, Woowoonga National Park (Qld). ©Hugh Nicholson

Fig. 530. Mixed dry sclerophyll forest-woodland, Mt Walsh NP (CQld). ©Geoff Williams

Fig. 531. View to the south from The Pinnacle, caldera rim, Border Ranges National Park (previously Wiangeree State Forest) (NNSW). ©Allen M. Sundholm

Fig. 532. Montane subtropical rainforest, Richmond Range (NNSW). ©Allen M. Sundholm

Fig. 533. Littoral rainforest margin (NNSW), habitat for *Maoraxia auroimpressa*, *Helferella miyal*, *Helferella manningensis* and *Paracupta lambertii*. ©Geoff Williams

Fig. 534. Coastal shrub complex, Crowdy Bay NP (NNSW). ©Geoff Williams

Fig. 535. Coastal rush – mangrove complex, Manning River estuary (NNSW), habitat for *Meliboeithon intermedium*. ©Geoff Williams

Fig. 538. Riparian forest complex, Barrington Tops (NNSW). ©Geoff Williams

Fig. 536. *Macrozamia* sp. cf. *communis* (Zamiaceae), Dingo Tops (NNSW). ©Geoff Williams

Fig. 539. Mixed flowering Faboideae genera in eucalypt-dominated sclerophyll forest (NNSW). ©Geoff Williams

Fig. 537. *Nothofagus moorei* (Nothofagaceae), larval host for *Nascioides tillyardi, Melobasis hypocrita,* cool temperate rainforest, Barrington Tops (NNSW). ©Geoff Williams

Fig. 540. *Gahnia* sp. cf. *sieberiana* (Cyperaceae), larval and adult host plant of *Synechocera deplana*. ©Geoff Williams

Fig. 541. *Kunzea ambigua* (Myrtaceae), Royal National Park (NSW). ©Geoff Williams

Fig. 544. Open woodland-shrub complex, Pilliga State Forest (CNSW). ©Geoff Williams

Fig. 542. *Angophora hispida* (Myrtaceae), Sydney (NSW). ©Geoff Williams

Fig. 545. Mixed eucalypt woodland, Warrumbungles National Park (WNSW). ©Geoff Williams

Fig. 543. *Leptospermum polygalifolium* (Myrtaceae), western Blue Mountains (NSW). ©Geoff Williams

Fig. 546. Box woodland (now rare due to vast clearing), (CNSW). ©Allen M. Sundholm

Fig. 547. Old growth *Acacia harpophylla* (Fabaceae) 'brigalow' woodland NNW of Brewarrina (NSW). ©Allen M. Sundholm

Fig. 550. Alpine heath-shrub-snow gum complex, Kosciuszko National Park (SNSW). ©Geoff Williams

Fig. 548. Semi arid grassy shrub complex, vicinity of Broken Hill (WNSW); habitat for *Australorhipis aphanochila*. ©Geoff Williams

Fig. 551. Temperate rainforest, Errinundra Plateau (NEVic). ©Hugh Nicholson

Fig. 549. Alpine woodland and shrub-tussock grassland complex, Kosciuszko National Park (SNSW). ©Geoff Williams

Fig. 552. Low closed shrub complex, Big Desert (NWVic). ©Geoff Williams

Fig. 553. Mallee–eucalypt woodland, vicinity of Tempy (NWVic). ©Geoff Williams

Fig. 554. Temperate rainforest, Otway Range (SWVic). ©Geoff Williams

Fig. 555. Woodland, Wilpena Pound (CSA). ©Allen M. Sundholm

Fig. 556. Mallee on white sand, Hincks Conservation Park (SWSA). ©Allen M. Sundholm

Fig. 557. Eucalypt-dominated 'mallee' woodland, Ellura Sanctuary east of Adelaide (SA). ©Brett and Marie Smith

Fig. 558. Mixed *Callitris gracilis* (Cupressaceae)-eucalypt woodland, Ellura Sanctuary east of Adelaide (SA). ©Brett and Marie Smith

Fig. 559. Summit heath, Ben Lomond (NETas). ©Allen M. Sundholm

Fig. 562. Open woodland, Keep River National Park (NWNT). ©Geoff Williams

Fig. 560. High altitude dense shrub complex, Hartz Mountains (Tas). ©Allen M. Sundholm

Fig. 563. Riverine closed forest, Mataranka (NNT), such habitats have been poorly investigated for Buprestidae. ©Geoff Williams

Fig. 561. Grassy open woodland, vicinity of Katherine (NNT). ©Geoff Williams

Fig. 564. Western Hammersley Ranges (CNT). ©Geoff Williams

Fig. 565. Flowering *Grevillea pterosperma* (Proteaceae), sandplain heath, Forrestania to The Breakaways (WA). ©Allen M. Sundholm

Fig. 568. *Acacia* (Mimosoideae) scrubland with *Senna* (Caesalpinioideae) in understorey, Gascoyne Region (NWWA). ©Allen M. Sundholm

Fig. 566. Flowering *Leptospermum erubescens* (Myrtaceae), sandplain heath, Forrestania to The Breakaways (WA). ©Allen M. Sundholm

Fig. 569. Acacia-dominated rock outcrop amongst eucalypt woodland, Canna Nature Reserve (WA). ©Allen M. Sundholm

Fig. 567. Flowering mallee-heathland, Mount Ragged, Cape Arid National Park (SWA). ©Allen M. Sundholm

Fig. 570. Woodland near mulga-eucalypt line, Bonnie Rock (WA). ©Allen M. Sundholm

Fig. 571. Root parasite *Nutysia floribunda* (Loranthaceae), restricted in range to SWWA. ©Allen M. Sundholm

Fig. 574. Mixed low closed heath-shrubland, Leeuwin-Naturaliste National Park (SWWA). ©Geoff Williams

Fig. 572. Coastal heath, Leeman area (SWWA). ©Allen M. Sundholm

Fig. 575. *Macrozamia*-eucalypt forest association, Shannon National Park (SWWA). ©Geoff Williams

Fig. 573. Flowering *Melaleuca uncinata* (Myrtaceae) in peak bloom amongst 'mallee' eucalypts (WA). ©Allen M. Sundholm

Fig. 576. Stirling Ranges (SWWA), extends east to west for 60 km and contains Bluff Knoll (1074 m), the highest eminence in southern Western Australia. ©Geoff Williams

4   Threats and conservation

Species with a higher degree of habitat specialisation, restricted distribution and whose populations exhibit low abundances, are especially vulnerable to anthropogenic-induced events and environmental change. Such species risk extinction (Reyes-Gonzáles et al. 2021).

Broadscale land clearing, as has happened in central New South Wales and Queensland, western Victoria, the Eyre Peninsula of South Australia, and the 'Wheatbelt' of Western Australia has resulted in the wholesale loss of the buprestid fauna, with relict bushland and roadside easements offering little opportunity for survival of diverse buprestid faunas (Sundholm and Catford 1982; Bellamy 2006). Likewise, wholesale loss of rainforest along the east coast and ongoing fragmentation of native heath, shrub, woodland communities and eucalypt-dominated forest progressively results in reduction of the distribution of species, isolation and extinction of populations, and species loss (Adam 1987; Turner and Hawkeswood 1996a; Williams 2020a, 2021).

Climate warming threatens plant communities, many components of which represent larval and adult buprestid food plants. Warmer temperatures can affect plant seed production and dispersal, leading to less seedling establishment and to vegetation dieback at larger scales, with a step-wise reduction in plant cover and possible loss of plant species. The various impacts of temperature change and altered rainfall can ultimately lead to a reduction or shift in the distribution of plant species, as well as broader changes in the composition of whole plant communities. Australia's tropical forests may be vulnerable if plants are unable to adjust to warming, their potential vulnerability being due to the adaptation over long periods of time of tropical plants to stable thermal conditions, unlike plants from cooler (higher latitudes) where seasonal temperatures fluctuate strongly (Crous 2019). Subsequent impacts on resident buprestid faunas may be tempered if food plant associations are 'plastic' (involving numerous plant species), but species with narrow host and habitat ranges have a heightened level of threat. Alpine heath, woodlands and coniferous rainforests, and their resident faunas, are at high risk due to increased dryness (leading to dieback) and greater threat from fire (Mallick 2013). Increases in ocean levels threaten coastal mangroves and saltmarsh. Owing to the conversion of much of the south-eastern coastline to residential and industrial areas, shipping facilities and 'hardened' encasement against erosion, there is limited opportunity for saltmarsh to colonise landward zones, although mangroves may find opportunity to colonise nearly inundated estuary strandlines. Consequently, populations of saltmarsh-associated species such as *Meliboeithon intermedium* may be increasingly at risk.

Fire constitutes an ever-present and growing threat, with landscape-scale fires, as occurred in eastern Australia during the summer of 2019–20, capable of completely removing larval and adult food plants; the capacity of the buprestid fauna to recover is unknown, threat examples from cataclysmic fires being the obligate larval associations of *Synechocera* with *Gahnia* (Cyperaceae) and *Xyroscelis* with Zamiaceae, and that of the Tasmanian endemic *Castiarina inscupta* and the associated plant *Ozothamnus hookeri* (Asteraceae). Less intense, but frequent, fires in remnant woodland and forest (which are vulnerable to complete incineration in single fire events) can result in inadequate time for plants dependent on regeneration from seed being able to mature and reproduce, leading to localised extinction of populations (Richards and Spencer 2020).

Alien plants and flower visitors (e.g. *Apis mellifera*) acting in concert can facilitate higher levels of invasion by exotic plants (Simpson *et al.* 2005), their impact potentially resulting in the smothering of native plants or the disruption to pollen transfer. Numerous alien vines, herbs and shrubs (as with Asparagaceae, *Chrysanthemoides monilifera*, *Anredera cordifolia*) are able to infest the canopies of lowland rainforest stands or smother the forest floor, disrupting ecological processes such as plant–pollinator interactions (Arroyo-Correa *et al.* 2019). The South American pathogen 'myrtle rust', *Austropuccinia psidii*, threatens Australian Myrtaceae (and consequently dependent anthophilous insects and vertebrates), with individual myrtaceous species, such as *Rhodomyrtus psidioides* (Williams 2018), exhibiting dramatic responses to its presence. A native to South America, and possibly Central America, myrtle rust was first discovered in New South Wales in 2010 and has since spread rapidly, reaching far North Queensland and the Northern Territory by 2015 (Makinson 2018). It is now widely established along the east coast, and has penetrated parts of the tableland in Queensland. The family Myrtaceae comprises about 10 per cent of Australia's native plant species, and is an important component of Australia's ecosystems. Myrtle rust has already proved capable of affecting more than 350 native plant species or subspecies in about 49 genera (Makinson 2018; Williams and Adam 2019), with the pathogen being potentially capable of initiating landscape-scale ecological impacts on native plant and animal communities, as well as the extinction of susceptible species. For example, Pegg *et al.* (2017) observed that in wet sclerophyll forest and rainforest sites in south-eastern Queensland there was a massive floristic and structural change with the prior 94 per cent dominance of Myrtaceae in the forest mid-storey being reduced to 63 per cent and 37 per cent in the 'regeneration' layer.

However, collecting can also constitute a threat. Buprestid beetles have long been a popular group, the large (*Calodema*, *Metaxymorpha*, *Temognatha*), colourful (*Stigmodera gratiosa*) and uncommon species being eagerly sought and often traded at high prices. The provision of collecting site details in research papers can serve as a catalyst for the seeking out of species, sometimes resulting in the suppression of the publication of precise locality records. Species occurring as scattered small populations, or relict populations in fragmented habitats, may be at particular risk, especially where these are repeatedly targeted. In Western Australia, owing to their appeal to collectors, all Buprestidae were gazetted as 'rare fauna' (see *Wildlife Conservation Act 1978*) and could not be collected without a scientific permit – though this listing 'did not prevent the almost simultaneous clearing of large patches of prime habitat for agricultural development in the state' (New 2007). It remains to be seen whether the Western Australian state government's superseding *Biodiversity Conservation Act, 2016* facilitates the conservation of buprestid species and the retention of their habitat at the landscape scale.

Appendix 1. List of buprestid genera recorded from Australia

(Subfamily placement follows Lawrence and Lemann 2019)
* = introduced taxon

SUBFAMILY POLYCESTINAE LACORDAIRE, 1857

Astraeus Laporte & Gory, 1837
Helferella Cobos, 1957
Paratrachys Saunders, 1873
Polycesta Seville, 1833
Prospheres Saunders, 1868
Strigoptera Dejean, 1833
Xyroscelis Saunders, 1868

SUBFAMILY CHRYSOCHROINAE LAPORTE, 1835

Austrochalcophora Bellamy, 2006
Austrophorella Kerremans, 1902–1903
Chalcophorotaenia Obenberger, 1928
Chrysodema Laporte & Gory, 1835
Cyphogastra Deyrolle, 1864
Iridotaenia Deyrolle, 1864
Metataenia Théry, 1923b
Paracupta Deyrolle, 1864
Pseudotaenia Kerremans, 1902–1903

SUBFAMILY BUPRESTINAE LEACH, 1815

Anilara Saunders, 1868
Anthaxoschema Obenberger, 1923
Araucariana Levey, 1978
Australorhipis Bellamy, 1986
Barakula Peterson, 2000
Belionota Eschscholtz, 1829
Bubastes Laporte & Gory, 1836
Buprestina Obenberger, 1923
**Buprestis* Linnaeus, 1758 (= *Cypriacis* Casey, 1909)
Burnsiellus Levey & Bellamy, 2013
Chrysobothris Eschscholtz, 1829

Cyrioides Carter, 1920
Diadoxus Saunders, 1868
Euryspilus Lacordaire, 1857
Julodimorpha Gemminger & Harold, 1869
Maoraxia Obenberger, 1937
Melobasis Gory & Laporte, 1837
Merimna Saunders, 1868
Microcastalia Heller, 1891
Nascio Laporte & Gory, 1837
Nascioides Kerremans, 1902–1903
Neobubastes Blackburn, 1892
Neobuprestis Kerremans, 1902–1903
Neocuris Saunders, 1868
Notobubastes Carter, 1924
Notographus Thomson, 1879
Pseudanilara Théry, 1911
Selagis Dejean, 1836
Theryaxia Carter, 1928
Torresita Gemminger & Harold, 1869

SUBFAMILY STIGMODERINAE SAUNDERS, 1871

Calodema Gory & Laporte, 1838
Calotemognatha Peterson, 1991
Castiarina Gory & Laporte, 1838
Metaxymorpha Parry, 1848
Stigmodera Eschscholtz, 1829
Temognatha Solier, 1933

SUBFAMILY AGRILINAE

Aaaaba Bellamy, 2013
Agrilus Curtis, 1825
Anthaxomorphus Deyrolle, 1865
Aphanisticus Latreille, 1829
Dinocephalia Obenberger, 1923
Diphucrania Dejean, 1833
Endelus Deyrolle, 1865

Ethonion Kubán, 2001
Germarica Blackburn, 1887
Habroloma Thomson, 1864
*Hedwigiella Obeneberger, 1941
Hypocisseis Thomson, 1879
Meliboeithon Obenberger, 1920
Neospades Blackburn, 1887
Pachycisseis Théry, 1929

Pachyschelus Solier, 1833
Paracephala Saunders, 1868
Sambus Deyrolle, 1865
Stanwatkinsius Barker & Bellamy, 2001
Synechocera Deyrolle, 1865
Toxoscelus Deyrolle, 1865
Trachys Fabricius, 1801

Appendix 2. Summary of larval and adult plant relationships

(After Bellamy *et al.* (2013); Bílý and Powell (2017); Lawrence and Lemann (2019); Nylander (2008); additional unpublished G. Williams records)
Key: 'adults' adults associated with leaves and/or flowers; 'larvae' larval host plant.

SUBFAMILY POLYCESTINAE LACORDAIRE, 1857
Astraeus
    Cupressaceae (*Callitris*), adults.
    Casuarinaceae (*Allocasuarina, Casuarina*), adults; (*Allocasuarina*), larvae.
    Fabaceae (*Acacia, Daviesia, Jacksonia*), adults.
    Myrtaceae (*Baeckea, Kunzea, Melaleuca, Micromyrtus*), adults; (*Eucalyptus*), larvae.
    Pittosporaceae/Bursariaceae (*Bursaria*), larvae.
    Proteaceae (*Banksia, Dryandra, Grevillea, Hakea*), adults; (*Banksia*), larvae.
    Xanthorrhoeaceae (*Xanthorrhoea*), adults.

Helferella
    Fabaceae, adults.
    Sapindaceae (*Guioa*), adults.

Paratrachys
    Moraceae (*Trophis*), adults.
    ?Moraceae (?*Ficus*), larvae.

Polycesta unknown

Prospheres
    Araucariaceae (*Araucaria*), larvae.
    Pinaceae (*Pinus*), larvae.

Strigoptera
    Bombacaceae (*Camptostemon*), larvae.
    Rhizophoraceae (*Ceriops*), adults.

Xyroscelis
    Zamiaceae (*Macrozamia*), adults, larvae.

SUBFAMILY CHRYSOCHROINAE LAPORTE, 1835
Austrochalcophora
    Fabaceae (*Acacia*), adults, larvae.

Austrophorella
    Rhamnaceae (*Alphitonia*), adults, larvae.

Chalcophorotaenia
    Anacardiaceae (*Pleiogynium*), adults.
    Fabaceae (*Gastrolobium*), larvae.
    Myrtaceae (*Eucalyptus, Melaleuca*), adults.
    Proteaceae (*Grevillea*), adults.

Chrysodema unknown

Cyphogastra
    Anacardiaceae (*Pleiogynium*), adults.
    Combretaceae (*Terminalia*), adults, larvae.
    Elaeocarpaceae (*Elaeocarpus*), larvae.
    Euphorbiaceae (*Aleurites*), larvae.
    Fabaceae (*Acacia*), adults, larvae.
    Lecythidaceae (*Barringtonia*), larvae; (*Planchonia*), adults.
    Rhamnaceae (*Alphitonia*), larvae.

Iridotaenia
    Euphorbiaceae (*Mallotus*), larvae.
    Myrtaceae (*Eucalyptus, Xanthostemon*), adults.

Metataenia Unknown

Paracupta Unknown

Pseudotaenia
    Casuarinaceae (*Casuarina*), adults.
    Fabaceae (*Acacia*), adults, larvae.
    Myrtaceae (*Eucalyptus*), adults.

SUBFAMILY BUPRESTINAE LEACH, 1815

Anilara
Casuarinaceae (*Allocasuarina*), adults.
Myrtaceae (*Eucalyptus, Leptospermum*), adults; (*Eucalyptus*), larvae.
Rutaceae/Flindersiaceae (*Flindersia*), larvae.

Anthaxoschema
Fabaceae (*Acacia*), larvae.

Araucariana
Araucariaceae (*Araucaria*), larvae.

Australorhipis unknown

Barakula unknown

Belionota
Anacardiaceae (*Mangifera*), larvae.
Bomabacaceae (*Ceiba*), adults, larvae.
Casuarinaceae (*Casuarina*), adults.
Fabaceae (*Delonix*), larvae.

Bubastes
Myrtaceae (*Melaleuca*), adults.

Buprestina unknown

Burnsiellus
Fabaceae (*Acacia*), larvae.
Myrtaceae (*Melaleuca*), larvae.

Chrysobothris
Casuarinaceae (*Allocasuarina*), larvae.
Fabaceae (*Acacia*), adults, larvae.
Myrtaceae (*Eucalyptus*), adults; (*Backhousia*), larvae.
Primulaceae (*Myrsine*), larvae.
Rhamnaceae (*Alphitonia*), larvae.

Cyrioides
Cunoniaceae (*Davidsonia*), larvae.
Myrtaceae (*Angophora, Leptospermum*), adults.
Proteaceae (*Banksia*), adults, larvae.

Diadoxus
Cupressaceae (*Callitris, Cupressus*), larvae.
Pinaceae (*Pinus*), larvae.

Casuarinaceae (*Allocasuarina*), adults, larvae.
Fabaceae (*Acacia*), larvae.
Proteaceae (*Grevillea*), larvae.

Euryspilus
Cyperaceae (undet. genus/species), adults.

Julodimorpha
Fabaceae (*Acacia*), adults.
Myrtaceae (*Calothamnus, Eucalyptus*), larvae.

Maoraxia
Podocarpaceae (*Podocarpus*), adults.
Celastraceae (*Elaeodendron*), adults, larvae.
Cunoniaceae (*Schizomeria*), adults.
Malvaceae (*Argyrodendron*), adults.
Myrtaceae (*Syzygium*), adults, larvae.
Rutaceae (*Acronychia*), adults.
Sapindaceae (*Alectryon, Guioa*), larvae.

Melobasis
Pinaceae (*Pinus*), larvae.
Asteraceae (*Ozothamnus*), adults.
Atherospermataceae (*Doryphora*), larvae.
Casuarinaceae (*Allocasuarina, Casuarina*), adults.
Cunoniaceae (*Cerapetalum*), larvae.
Cyperaceae (*Gahnia*), adults.
Ericaceae (*Leucopogon*), adults.
Euphorbiaceae (*Excoecaria*), larvae.
Fabaceae (*Acacia, Aotus, Bossiaea, Cassia, Chorizema, Daviesia, Dillwynia, Gastrolobium, Gompholobium, Jacksonia, Mirbelia, Otion, Phyllota, Platypodium, Pultenaea, Viminaria*), adults; (*Acacia, Bossiaea, Jacksonia Oxylobium, Pultenaea, Urodon, Viminaria*), larvae.
Lamiaceae (*Clerodendrum*), larvae.
Loranthaceae (*Nuytsia*), adults.
Malvaceae (*Triumfetta*), adults.
Myrtaceae (*Baeckea, Corymbia, Eucalyptus, Leptospermum, Melaleuca, Syzygium*), adults; (*Corymbia*), larvae.
Myrsinaceae (*Aegiceras*), larvae.
Nothofagaceae (*Nothofagus*), larvae.
Pittosporaceae (*Pittosporum*), larvae, adults.
Poaceae (*Themeda*), adults.
Proteaceae (*Adenanthos, Banksia, Hakea*), adults
Rhamnaceae (*Alphitonia*), adults, (*Alphitonia*), larvae.
Rutaceae (*Citrus, Flindersia, Rhadinothamnus*), larvae.

Santalaceae (*Santalum*), adults.
Scrophulariaceae (*Myoporum*), adults.
Xanthorrhoeaceae (*Xanthorrhoea*), adults.

Merimna

Myrtaceae (*Angophora*), adults; (*Baeckea, Corymbia, Eucalyptus, Melaleuca*), larvae.

Microcastalia unknown

Nascio

Myrtaceae (*Eucalyptus*), adults; (*Corymbia, Eucalyptus, Metrosideros*), larvae.

Nascioides

Asteraceae (*Taraxacum*), adults.
Atherospermataceae (*Doryphora*), larvae.
Cunoniaceae (*Ceratopetalum, Schizomeria*), larvae.
Elaeocarpaceae (*Sloanea*), larvae.
Fabaceae (*Acacia*), adults.
Malvaceae (*Argyrodendron*), larvae.
Myrtaceae (*Angophora, Corymbia, Leptospermun, Syzygium*), adults; (*Backhousia, Eucalyptus, Leptospermum, Syzygium*), larvae.
Nothofagaceae (*Nothofagus*), adults, larvae.
Rosaceae (*Crataegus*), adults.
Rutaceae) (*Flindersia*), larvae.

Neobubastes

Casuarinaceae (*Allocasuarina*), adults.
Fabaceae (*Acacia*), adults.
Myrtaceae (*Thryptomene* or *Baeckia*), adults; (*Melaleuca*), larvae.

Neobuprestis

Fabaceae (*Acacia*), adults.

Neocuris

Anacardiaceae (*Euroschinus*), adults.
Apiaceae (*Actinotus*), adults.
Asteraceae (*Taraxacum*), adults.
Epacridaceae (*Epacris*), adults.
Fabaceae (*Pultenaea*), larvae.
Loranthaceae (*Nuytsia*), adults.
Myrtaceae (*Angophora, Backhousia, Baeckea, Kunzea, Leptospermum, Syzygium*), adults.
Pittosporaceae/Bursariaceae (*Bursaria*), adults.
Proteaceae (*Conospermum*), adults.

Notobubastes

Fabaceae (*Acacia*), adults.

Notographus

Malvaceae (*Argyrodendron*), larvae.

Pseudanilara

Casuarinaceae (*Casuarina*), adults.
Cunoniaceae (*Ceratopetalum*), larvae.
Myrtaceae (*Corymbia, Eucalyptus, Syzygium*), adults; (*Eucalyptus, Kunzea, Rhodomyrtus, Syzygium*), larvae.

Selagis

Casuarinaceae (*Allocasuarina/Casuarina*), larvae.
Myrtaceae (*Angophora, Baeckea, Corymbia, Eucalyptus, Kunzea, Leptospermum, Syzygium*), adults; (*Eucalyptus*), larvae.
Pittosporaceae/Bursariaceae (*Bursaria*), adults.

Theryaxia

Cupressaceae (*Callitris*), adults, larvae.

Torresita

Cunoniaceae (*Ceratopetalum*), adults.
Myrtaceae (*Angophora, Eucalyptus, Kunzea, Leptospermum, Melaleuca, Syzygium, Melaleuca*), adults; (*Rhodomyrtus*), larvae.
Rhamnaceae (*Alphitonia*), adults.

SUBFAMILY STIGMODERINAE SAUNDERS, 1871
Calodema

Elaeocarpaceae (*Elaeocarpus*), adults.
Fabaceae (*Bauhinia*), adults.
Myrtaceae (*Corymbia, Leptospermum, Syzygium, Tristaniopsis*), adults.
Proteaceae (*Buckinghamia, Macadamia*), adults.
Rousseaceae (*Cuttsia*), adults.
Rutaceae (*Melicope*), adults.

Calotemognatha

Loranthaceae (*Nuytsia*), adults.
Myrtaceae (*Eucalyptus, Melaleuca*), adults.

Castiarina

Acanthaceae (*Avicennia*), adults.
Anacardiaceae (*Euroschinus*), adults.
Apiaceae (*Daucus*), adults.

Asteraceae (*Cassinia*, '*Helichrysum*', *Olearia*, *Ozothamnus*), adults.

Cannabaceae (*Aphananthe*), adults.

Cunoniaceae (*Ceratopetalum*, *Karrabina*), adults.

Epacridaceae (*Epacris*), adults.

Fabaceae (*Acacia*, *Barklya*, *Dillwynia*, *Jacksonia*, *Phyllota*), adults.

Lamiaceae (*Clerodendrum*, *Dicrastylis*), larvae.

Loranthaceae (*Muellerina*, *Nuytsia*), adults; (*Muellerina*), larvae.

Myrtaceae (*Angophora*, *Baeckea*, *Callistemon*, *Calytrix*, *Chamelaucium*, *Corymbia*, *Eucalyptus*, *Harmogia*, *Hypocalymma*, *Kunzea*, *Leptospermum*, *Melaleuca*, *Micromyrtus*, *Scholtzia*, *Syzygium*, *Thryptomene*, *Tristaniopsis*, *Verticordia*), adults; (*Eucalyptus*), larvae.

Pittosporaceae/Bursariaceae (*Bursaria*), adults.

Proteaceae (*Buckinghamia*, *Cardwellia*, *Conospermum*, *Darlingia*, *Grevillea*, *Hakea*, *Macadamia*), adults.

Putranjivaceae/Euphorbiaceae (*Drypetes*), adults.

Rhamnaceae (*Alphitonia*), adults.

Rousseaceae (*Cuttsia*), adults.

Rubiaceae (*Psydrax*), adults.

Rutaceae (*Acronychia*, *Eriostemon*, *Melicope*), adults.

Salicaceae (*Homalium*, *Scolopia*), adults.

Sapindaceae (*Guioa*), adults; (*Dodonaea*), larvae.

Scrophulariaceae (*Calamphoreus*, *Eremophila*, *Myoporum*), adults.

Thymelaeaceae (*Pimelia*), adults.

Xanthorrhoeaceae (*Xanthorrhoea*), adults.

Metaxymorpha

Cannabaceae (*Aphananthe*), adults.

Casuarinaceae (*Casuarina*), adults, larvae.

Elaeocarpaceae (*Elaeocarpus*), adults.

Myrtaceae (*Corymbia*, *Eucalyptus*, *Leptospermum*, *Syzygium*, *Tristaniopsis*), adults.

Proteaceae (*Buckinghamia*, *Grevillea*, *Macadamia*), adults.

Rubiaceae (*Canthium*, *Cyclophyllum*, *Psydrax*), adults; (*Cyclophyllum*), larvae.

Rutaceae (*Melicope*), adults.

Sapindaceae (*Guioa*, *Sarcopteryx*, *Synima*), adults; (*Guioa*, *Sarcopteryx*), larvae.

Stigmodera

Loranthaceae (*Nutsyia*), adults.

Myrtaceae (*Agonis*, *Chamelaucium*, *Leptospermum*, *Melaleuca*), larvae; (*Angophora*, *Baeckea*, *Eucalyptus*, *Kunzea*, *Leptospermum*), adults.

Temognatha

Apiaceae (*Actinotus*), adults.

Asteraceae (*Cassinia*), adults.

Casuarinaeae (*Allocasuarina*, *Casuarina*), larvae.

Loranthaceae (*Nuytsia*), larvae.

Myrtaceae (*Angophora*, *Baeckea*, *Corymbia*, *Eucalyptus*, *Leptospermum*, *Melaleuca*), adults; (*Corymbia*, *Eucalyptus*), larvae.

Pittosporaceae/Bursariaceae (*Bursaria*), adults.

Proteaceae (*Grevillea*), adults.

Scrophulariaceae (*Myoporum*), adults.

Xanthorrhoeaceae (*Xanthorrhoea*), adults.

SUBFAMILY AGRILINAE

Aaaaba

Rosaceae (*Rubus*), adults, larvae.

Agrilus

Casuarinaceae (*Allocasuarina*), adults; (*Allocasuarina*), larvae.

Clusiaceae (*Hypericum*), larvae.

Euphorbiaceae (*Claoxylon*), larvae.

Fabaceae (*Acacia*), adults; (*Acacia*), larvae.

Malvaceae (*Argyrodendron*, *Hibiscus*), larvae.

Myrtaceae (*Eucalyptus*, *Leptospermum*), larvae.

Putranjivaceae (*Drypetes*), larvae.

Rutaceae (*Citrus*), adults, larvae.

Anthaxomorphus unknown

Aphanisticus unknown

Dinocephalia

Casuarinaceae (*Allocasuarina*, *Casuarina*), adults; (*Allocasuarina*), larvae.

Fabaceae (*Dillwynia*), adults.

Myrtaceae (*Leptospermum*), adults.

Pittosporaceae/Bursariaceae (*Bursaria*), adults.

Diphucrania

Cupressaceae (*Callitris*), adults.

Casuarinaceae (*Allocasuarina*, *Casuarina*), adults.

Epacridaceae (*Epacris*), adults.
Ericaceae (*Leucopogon*), adults.
Fabaceae (*Acacia, Dillwynia, Goodenia, Jacksonia, Phyllota, Pultenaea, Tephrosia Viminaria*), adults; (*Acacia, Bossiaea, Dillwynia, Pultenaea*), larvae.
Haloragaceae (*Glischrocaryon*), adults.
Iridaceae (*Patersonia*), adults.
Myrtaceae (*Angophora, Baeckea, Chamelaucium, Eucalyptus, Kunzea, Leptospermum, Melaleuca, Syzygium*), adults.
Pittosporaceae/Bursariaceae (*Bursaria*), adults.
Proteaceae (*Banksia, Grevillea*), adults.
Sapindaceae (*Dodonaea*), adults, (*Elattostachys*), larvae.
Xanthorrhoeaceae (*Xanthorrhoea*), adults, larvae.
Xyridaceae (*Xyris*), adults.

Endelus unknown

Ethonion
Fabaceae (*Daviesia, Dillwynia, Jacksonia, Leptospermum, Phyllota, Pultenaea*), adults; (*Dillwynia, Pultenaea*), larvae.
Iridaceae (*Patersonia*), adults.

Germarica
Casuarinaceae (*Allocasuarina, Casuarina*), adults.

Habroloma
Elaeocarpaceae (*Elaeocarpus*), adults.
Fabaceae (*Cantharospermum, Desmodium, Kennedia*), adults; (*Kennedia*), larvae.

Hypocisseis
Casuarinaceae (*Casuarina*), adults.
Celastraceae (*Maytenus*), adults.
Loranthaceae (*Amyema*), adults.
Myrtaceae (*Leptospermum*), adults.
Proteaceae (*Hakea*), adults.
Rhamnaceae (*Alphitonia*), adults.
Santalaceae (*Exocarpus*), adults, larvae.

Meliboeithon
Casuarinaceae (*Casuarina*), adults.

Cyperaceae (*Isolepis*), adults.
Fabaceae (*Oxylobium*), ?larvae.
Juncaceae (*Juncus*), adults.

Neospades
Ebenaceae (*Diospyros*), ?larvae.
Fabaceae (*Acacia, Atylosia, Cantharospermum, Jacksonia, Tephrosia*), adults.
Malvaceae (*Hibiscus, Sida*), adults.
Myrtaceae (*Corymbia, Eucalyptus*), adults.
Solanaceae (*Solanum*), adults.

Pachycisseis
Fabaceae (*Acacia*), adults.
Myrtaceae (*Leptospermum*), adults.
Grevillea (Proteaceae), adults.

Pachyschelus
*Cecropiaceae, larvae.
*Euphorbiaceae (*Croton*), larvae.
*Fabaceae (undet. genus/species), larvae.
* = foreign host records.

Paracephala
Casuarinaceae (*Allocasuarina*), adults, larvae.
Poaceae (*Chrysopogon, Stipa*), adults.

Sambus unknown

Stanwatkinsius
Casuarinaceae (*Allocasuarina, Casuarina*), adults.
Proteaceae (*Banksia, Grevillea, Hakea*), adults.

Synechocera
Cyperaceae (*Gahnia*), adults, larvae.
Fabaceae (*Acacia*), adults.
Myrtaceae (*Leptospermum*), adults.
Xanthorrhoeaceae (*Xanthorrhoea*), adults.

Toxoscelus unknown

Trachys unknown

Appendix 3. Pollen loads from Buprestidae collected in lowland subtropical rainforest and wet sclerophyll forest

(From Williams and Adam 1998)

Key: Multiple listings of individual species indicate differences in collection sites cited in original reference. Approximate pollen loads given in three abundance classes; 'M' moderate = >100 grains; 'H' Heavy + >200 grains; 'A' abundant = >300 grains; total count given where <100 grains carried. '>99' = probably all 'Home' pollen, home pollen being that of the plant upon which the beetle was collected.

Buprestid species	Plant species from which the beetle was collected	Pollen load	% Home pollen
Castiarina acuminata	Alphitonia excelsa (Rhamnaceae)	6	100
C. acuminata	Alphitonia excelsa	54	98
C. acuminata	Cuttsia viburnea (Rousseaceae)	A	<30
C. acuminata	Guioa semiglauca (Sapindaceae)	M	>95
C. acuminata	Guioa semiglauca	A	>99
C. delta	Tristaniopsis laurina (Myrtaceae)	A	>99
C. insignis	Syzygium smithii (Myrtaceae)	A	100
C. insignis	Tristaniopsis laurina	H	>99
C. insignis	Syzygium floribundum	A	100
C. neglecta	Alphitonia excelsa	M	100
C. neglecta	Alphitonia excelsa	M	?100
C. producta	Syzygium smithii	A	100
C. producta	Alphitonia excelsa	M	Unrecorded
C. producta	Guioa semiglauca	A	>70
C. producta	Tristaniopsis laurina	A	<70
C. producta	Tristaniopsis laurina	M	?100
C. producta	Syzygium floribundum	A	100
C. pulchripes	Cuttsia viburnea	A	>95
C. ?sexcavata	Syzygium smithii	A	>95
C. ?vicina	Syzygium floribundum	A	>99
Metaxymorpha grayii	Tristaniopsis laurina	A	>99
Selagis aurifera	Syzygium floribundum	A	>70
S. splendens	Syzygium smithii	M	>50
S. splendens	Syzygium floribundum	A	>99
Torresita cuprifera	Syzygium floribundum	M	>95

Appendix 4. Early taxonomists and collectors: 1770–1950

The history of European engagement with the Australian beetle fauna began with the collections made by Joseph Banks in 1770 during the first voyage of exploration undertaken by James Cook. However, as Barker (2006a) points out Cook's arrival, and the efforts of his naturalists, were mistimed for they arrived on the eastern seaboard at a time when adult buprestids were unlikely to have emerged (April–August). The collection of the first of Australia's rich buprestid fauna would have to wait some years.

Here we briefly note many (and cite examples of taxa) of those who worked on specimens sent to them, or who had the good fortune to collect and study first hand. It is by no means an exhaustive list and the 'cut off' point for those cited is an arbitrary one. Later entomologists and collectors are discussed in Barker (2006a), or are mentioned in the preceding text. Among the taxonomists mentioned by us, the contributions of Thomas Blackburn, Herbert Carter, Hippolyte Gory, Charles Kerremans, Francois Laporte, William Macleay, Jan Obenberger, Edward Saunders, Charles Waterhouse and John Westwood stand prominent.

Baly, Joseph Sugar (1816–90). Born in Warwick, England and died there. Baly was a Doctor of Medicine with an additional interest in Coleoptera, especially Chrysomelidae. His collection is in the British Museum. Among the material he acquired, presumably from contacts in Australia, was a Victorian buprestid specimen of what Levey and Bellamy (2013) were to describe as *Burnsiellus lobatum*.

Barnard, George (1830–94). Born in Chislehurst, England and died in Launceston, Tasmania. An entomologist and ornithologist, he lived near Duaringa, Central Queensland. His private collection went to the Tring Museum, England. Barnard wrote a paper '*Chalcophora* in the scrubs of Central Queensland', which was published in *The Entomologist*, 1890. His son Wilfred Bourne Barnard also became seriously interested in entomology and collected extensively in Queensland, including Cape York, northern New South Wales and Western Australia.

Blackburn, Thomas (Rev. Canon) (1844–1912). Born in Islington, England and died at Woodville near Adelaide. Blackburn was a Church of England clergyman who came to Australia in 1882. He was appointed an Honorary Curator at the South Australian Museum. During his time in Australia Blackburn described more than 3000 species of Australian Coleoptera, the majority of his type specimens being placed in the British Museum. His publications included descriptions of new species of *Bubastes*, *Burnsiellus* (as *Strigoptera*), *Castiarina*, *Chalcophora*, *Cyrioides* (as *Cyria*), *Diadoxus*, *Dinocephalia* (as *Paracephala*), *Ethonion* (as *Ethon*), *Iridotaenia* (as *Paracupta*), *Melobasis*, *Metaxymorpha*, *Neobubastes*, *Neocuris*, *Neobuprestis* (as *Strigoptera*), *Neospades*, *Paracephala* and *Selagis* (as *Curis*). Included among these was *Hypostigmodera variegata*, an unusual species in which the males possess distinctive strongly biflabellate antennae, but those of the female are normally serrate. Numerous species are named in his honour (see Bellamy 2002).

Boheman, Charles Henri (1796–1868). Born Jönköping, Sweden and Intendant at the Reichsmuseum, Stockholm. Described *Ethonion corpulentum* (as *Buprestis*).

Boisduval, Jean Baptiste Alphonse Dechauffour de (1799–1879). Doctor of Medicine and eminent entomologist. He described insects (including Buprestidae) collected on the voyage of discovery to the Pacific by the *Astrolabe* (1826–29). Buprestidae described by him included *Castiarina erythroptera*, *Cyrioides australis*, *Nascio vetusta* and *Selagis caloptera* (all as *Buprestis*). *Bubastes boisduvali* was named in his honour.

Brown, Horace (1883–1958). An electrical engineer and collector. He travelled and collected extensively in Australia. *Castiarina browni*, *Diphucrania browni*, *Paracephala browni* and numerous other species are named in his honour.

Brown, Robert (1773–1858). Born in Montrose, Scotland, and died in London. He was a botanist and naturalist on board the *Investigator*, commanded by Matthew Flinders. Brown was in Australia from 1801 until 1805, and visited islands in Bass Strait, settlements in Tasmania, and the Hawkesbury, Hunter and Williams River valleys in New South Wales. His insect collections were described by W. S. Macleay and William Leach, with specimens of *Castiarina* being described by William Kirby.

Carter, Herbert James (1858–1940). Born in Marlborough, England, taught at Sydney Grammar School and later as the Principal of Ascham School for Girls (Edgecliff) until 1914. Honorary Entomologist at the Australian Museum and collected in eastern Australia, and South and Western Australia. Published on several Coleoptera groups, especially Buprestidae and Tenebrionidae, the former including revisions and descriptions of *Anilara, Aphanisticus, Castiarina, Chrysobothris, Cisseis, Curis, Cyria, Cyrioides, Ethon, Melobasis, Neobuprestis, Neocuris, Neospades, Notobubastes, Paracephala, Pseudanilara, Stigmodera, Synechocera, Temognatha, Theryaxia* and *Trachys*; many of these (e.g. *Cisseis, Curis, Cyria, Ethon, Paracephala*) now transferred to different genera. In 1929 Carter published a checklist of the Australian Buprestidae, this appearing in the *Proceedings of the Linnean Society of New South Wales*. In 1933 he published *Gulliver in the Bush* (Angus and Robertson), a fascinating book documenting his many adventures in search of beetles. Carter's private collections were placed in the Museum of Victoria, Melbourne, the Australian Museum, Sydney, and the Australian National Insect Collection, Canberra.

Chevrolat, Auguste (1799–1884). French coleopterist who lived in Paris. Published a number of descriptions of beetles including Curculionidae, Cerambycidae and Cleridae, but also described *Ethonion reichei* (as *Diphucrania*) and *Stigmodera gratiosa*.

Chisholm, Edwin Claude (1880–1944). Doctor of Medicine and naturalist. Chisholm practised at Marrangee, Comboyne, Croydon and Mosman, New South Wales. Died at Blaxland. Published several papers in the *Australian Zoologist* on the fauna of the County of Cook (New South Wales), but in particular the fauna of the Comboyne Plateau (now largely cleared of rainforest). Corresponded with a number of Australian entomologists.

Clark, Charles Bagnall. Medical Superintendent at Prince Alfred Hospital, Sydney. Naturalist with interests in butterflies and beetles. Field collecting included the Blue Mountains and environs. An early collecting influence on Herbert Carter.

Cunningham, Allan (1791–1839). Born in Wimbleton, England and died in the Sydney Botanic Gardens, 27 June 1839. Though a botanist, during his extensive travels Cunningham also collected insects, the descriptions of these being published by J. Boisduval and W. S. Macleay, among others.

Dejean, Pierre Francois Marie Auguste (Count) (1780–1845). His descriptions of Australian Buprestidae included species described under *Strigoptera*; however, many of his names became invalid. He amassed a vast collection of Coleoptera (>20 000 species), and rose to the rank of Lieutenant General, and was *aide de campe* to Napolean, during the Napoleonic Wars.

Deuquet, Camille Felix (1882–1972). Teacher and entomologist. Deuquet was born in Belgium but at the start of World War I moved to England, then Australia. Here he joined the New South Wales Department of Education as a teacher in 1916, but enlisted in the Australian Infantry Force and served overseas during 1917–18 (during which he was wounded). After the war Deuquet returned to the Department of Education, and collected at various locations in New South Wales. He returned twice to Belgium, dying there at the age of 90. His collections are lodged in the Royal Museum, Brussels, with a smaller collection in the Museum of Nantes. Deuquet described a number of *Castiarina*. *Castiarina deuqueti* was described by H. J. Carter in his honour.

Deyrolle, Achille (1813–65). French entomologist mainly interested in Coleoptera. He owned a taxidermy and natural history shop in Paris. His descriptions of Australian Buprestidae include *Aaaaba nodosus* (as *Alcinous*), *Agrilus marmoreus, Hypocisseis brachyformis* (as *Cisseis*) and *Melobasis maculata* (as *Diceropygus*). *Castiarina deyrollei* and *Agrilus deyrollei* are named in his honour.

Dodd, Frederick Parkhurst (1861–1937). Naturalist and entomological collector with particular interests in butterflies and beetles. Born in Victoria but settled in Kuranda, North Queensland, becoming known as the 'Butterfly Man of Kuranda'. *Castiarina doddi* and *Agrilus doddi* are named in his honour. His son, Alan Parkhurst Dodd (1896–1981), was an entomologist with the Queensland Government in charge of the prickly pear control investigations, but like his father maintained an interest in collecting buprestids.

Donovan, Edward (1768–1837). English entomologist and natural history painter, and founder of the London Museum and Institute of Natural History (1807). Described *Castiarina undulata, Melobasis splendida, Stigmodera cancellata* (all as *Buprestis*) and *Temognatha grandis* (as *Stigmodera*). *Temognatha donovani* was named in his honour.

du Boulay, Francis Houssemayne (1837–1914). Collector, but also a talented entomological illustrator. Born in Kent, England, and arrived in Australia about 1857. Collected insects, mostly Coleoptera, in Western Australia and later extensively in eastern Australia. Finally settled at Beechworth, Victoria. *Temognatha duboulayi* was named in his honour. His two sons, William and Edgar, also became keen collectors.

Erichson, Wilhelm Ferdinand (1809–1849). German entomologist, trained as a medical doctor. Born in Stralsund on the Baltic. Described *Melobasis hypocrita*, *M. prisca*.

Fabricius, Johann Christian (1745–1810). Born in Schleswig, Denmark. He was the first entomologist to name insects from Australia. Fabricius was a pupil of the Swedish naturalist Linné (Linnaeus) and was entrusted with the role of describing the specimens collected during Cook's exploration. Examples of the buprestids he named include *Cyphogastra farinosa*, *Cyrioides imperialis*, *Diphucrania duodecimmaculata*, *Lampetis fastuosa*, *Melobasis purpurascens* and *Synechocera deplana* (all originally as *Buprestis*).

Fairmaire, Léon (1820–1906). Born in Paris, educated in law and spent 30 years in the Institut National du Service Public. His collection is in the Paris Museum. Described *Agrilus indigenus* as well as numerous *Selagis* (as *Curis*) and *Neocuris*.

French, Charles Hamilton (1868–1950). Botanical collector, horticulturalist, naturalist and entomological collector. Government Entomologist of Victoria. Published on insect pests but also collected insects for his father Charles French (1843–1933). Charles French (snr) was Assistant Government Entomologist (Victoria) and later biologist to the Victorian Department of Agriculture. The Coleoptera collections made by Charles French (snr) went to Neervort van de Poll, G. J van Lansberge and Museum Victoria.

Froggatt, Walter Wilson (1858–1937). Published prodigiously on Australian insects. Born in Victoria. Died at Croydon, New South Wales. In 1885 appointed assistant zoologist and entomologist to the Royal Geographical Society's New Guinea Expedition, and later collected in North Queensland for the Macleay Museum. In 1896, appointed Government Entomologist to the New South Wales Department of Agriculture but upon retirement in 1923 he was appointed Forest Entomologist to the Forestry Commission of New South Wales.

Géhin, Joseph James Baptiste (1816–89). French apothecary in Metz, naturalist and entomologist. Specialised in Coleoptera and Diptera. Sold his collection to René Oberthür. Described *Calotemognatha yarelli elegans* (as *Themognatha elegans*), *C. yarelli flavipennis* (as *Themognatha flavipennis*), *Temognatha chevrolatii* and *T. stevensii*.

Gemminger, Max (1820–87). German entomologist. In cooperation with (Baron) Edgar Harold, published the *Catalogus Coleopterorum*; completed in 1876. Volume V (1869) included Buprestidae.

Germar, Ernst Friedrich (1786–1853). Born in Glauchau, Sachsen. Professor of Minerology at Halle but also interested in entomology. Described *Amorphosoma tasmanicum*, *Diphucrania notulata* (as *Ethon*), *Melobasis simplex* (as *Buprestis*) and *Neospades chrysopygia* (as *Coraebus*). *Bubastes germari* and *Germarica* named in his honour.

Gestro, Raffaello (1845–1925). Distinguished Italian Coleopterist. Born in Genoa and was the Director of the Natural History Museum of Giacomo Doria. Described *Diphucrania albertisii* (as *Cisseis*), *Habroloma australasiae* (as *Trachys*) and *Neospades cuprifera* (as *Cisseis*).

Goerling, Hermann Emil August (1865–1941). Collector and exporter of native animals. Goerling was born in Geestemunde, Germany. He migrated to Australia, moving first to Melbourne, then Gippsland. He returned for a period to Germany before again migrating to Australia, this time settling in Western Australia. *Astraeus goerlingi*, *Castiarina goerlingi*, *Cisseis augustgoerlingi*, *Cisseis goerlingi* (a synonym of *Cisseis cyaneopyga*), *Cisseis* now a synonym of *Diphucrania*, and *Melobasis goerlingi* were named in his honour.

Gooding, Charles George Llwellen (1896–1980). Entomologist who lived in Victoria. Interested in Lepidoptera but also amassed a large collection of buprestids. Collection donated to the Australian National Insect Collection. *Castiarina goodingi* was named in his honour.

Gory, Hippolyte Louis (1800–52). French entomologist born in Paris. Publications, often jointly with Francois Laporte, included descriptions of *Calodema regale* (as *Stigmodera*), *Castiarina ocelligera* (as *Stigmodera*), *Calotemognatha yarelli* (as *Themognatha*), *Diphucrania marmorata* (as *Cisseis*), *Nascio xanthura*, *Selagis aurifera*. *Temognatha goryi* was named in his honour.

Gray, George Robert (1808–72). Born in London, England. Zoologist, ornithologist and author. Head of the ornithological section of the Natural History Museum, London. Published *Entomology in Australia* (1883). His buprestid specimens were described by Adam White. *Metaxymorpha grayii* was named in his honour.

Hacker, Henry Pollard (1876–1973). Born in Essex, England, and joined the staff of the Queensland Museum in 1911 as Entomologist. Hacker collected in Central Australia and Queensland. *Trachys hackeri* (= *Habroloma frenchi*) and *Castiarina hackeri* (=*C. rollei*) were named in his honour.

Harold, Edgar von (Baron) (1830–86). Born in Munich, Bavaria. Officer in the German army but wrote a number of papers on Coleoptera, cooperating with Dr Max Gemminger on the publication of the *Catalogus Coleopterorum*, completed in 1876. Described *Castiarina gravis* (as *Stigmodera*). *Castiarina haroldii* was named in his honour.

Haswell, William Aitcheson (1854–1925). Born in Edinburgh, Scotland. Appointed Curator of the Queensland Museum (1880), Acting Curator, Australian Museum (1883), Professor of Biology (1890–1915) and later Professor of Zoology (1915–18) at the University of Sydney. *Castiarina haswelli* was named in his honour.

Helms, Richard (1858–1914). Born in Germany and arrived in Australia in 1858. His collection of Australian and New Zealand Coleoptera is in the Bernice Pauahi Bishop Museum, Honolulu. He wrote a number of accounts of the Australian alpine insect fauna. *Castiarina helmsi* and *Cyrioides imperialis helmsi* were named in his honour.

Hope, Frederick William (1797–1862). Born in London, England. Graduated from Oxford University and presented his collections to that university in 1849, endowing it with a Professorship of Zoology. Published several papers on the Buprestidae, Cerambycidae, Lucanidae and others. His privately published pamphlet 'Buprestidae' (1836), later called 'Synopsis of Australian Buprestidae', has been the cause of controversy over the authorship of names subsequently published in later catalogues (discussed in Bellamy 2002). Hope's species included *Agrilus aurovittatus*, *Anilara adelaidae* (as *Anthaxia*), *Castiarina adelaidae* (as *Stigmodera*), *Chrysobothris australasiae*, *Diphucrania roseocuprea* (as *Ethon*), *Nascioides parryi* (as *Stigmodera*), *Neocuris fortnumii* (as *Anthaxia*), *Iridotaenia albivittis* (as *Buprestis*), *Paracephala pistacina* (as *Agrilus*) and *Stigmodera sanguinosa*. Several buprestid species (e.g. *Selagis hopei*) were named in his honour.

Illidge, R (1846–1929). Naturalist, born in London, England. Pioneer entomologist, who collected extensively in south-east Queensland. *Melobasis illidgei* is named in his honour.

Jordan, Karl (1861–1959). German–British entomologist, born at Hannover and educated at Göttingen University. Curator of the Entomological Department, Zoological Museum, Tring, England. Published over 400 papers. Described *Calodema plebeium*.

Kerremans, Charles (1847–1915). Born in 1847, and died in Brussels, Belgium. Kerremans specialised in the Buprestidae and is particularly known for his *Monographie des Buprestides*, published in several parts between 1904 and 1913. His first collection went to the British Museum, but his second collection was deposited in the Paris Museum. Species described included *Agrilus deyrollei*, *Castiarina acuminata* (as *Stigmodera*), *Chrysobothris regina*, *Endelus subcornutus*, *Meliboeithon crassum* (as *Paracephala*), *Pseudotaenia cerata* (as *Chalcotaenia*), *Temognatha aestimata* (as *Stigmodera*), *Torresita parallela* and *Trachys blackburni*; as well as several species of *Melobasis*. *Castiarina kerremansi* and *Pseudanilara kerremansi* were named in his honour.

Kershaw, James Andrew (1866–1946). Curator of the Zoological Collections, National Museum, Melbourne, 1899–1913, Director (1930–31). Published on Lepidoptera and with others contributed in 1908 to the reservation of Wilson's Promontory (southern Victoria) as a national park and sanctuary for flora and fauna (serving as Honorary Secretary on its management committee 1908–46). *Castiarina kershawi* was named in his honour.

Kirby, William (Rev.) (1759–1850). Born in Suffolk, England, and collaborated with William Spence on the *Introduction to Entomology*. His specimens are in the British Museum. *Bubastes kirbyi* and *Castiarina kirbyi* were named in his honour. Species described by Kirby include *Castiarina decemmaculata* and *Ethonion fissiceps* (both as Buprestis), and *Melobasis cupriceps* and *Torresita cuprifera*. *Bubastes kirbyi* and *Castiarina kirbyi* were named in his honour.

Krefft, Johann Louis Gerhard (1830–80). Born in Brunswick, Germany, and went to Melbourne in 1852, then to the Victorian gold diggings where he worked until 1857. Was later engaged as an assistant at the Victorian Museum. He briefly returned to Germany, but eventually returned to Australia and took up a number of positions with the Australian Museum, becoming Curator in 1860 until resigning in 1874. *Castiarina kreffti* recalls his name.

Lacordaire, Jean Théodore (1801–70). Born in Recey-sur-Ource, France, and died at Liège, Belgium. Professor of Zoology and later of Comparative Anatomy at Liège. Lacordaire was considered to be among the outstanding Coleopterists of his time. Studies included that of *Euryspilus*, Polycestinae and Stigmoderini.

Laporte, Francois Louis Nompar de Caumont (Comte de Castelnau) (1810–80). Born in London, England and died in East Melbourne, Australia. Filled official posts in French colonies and was noted as an ichthyologist and entomologist. He wrote under the name of Laporte or Delaporte, and later Castelnau. Papers were sometimes co-authored with H. L. Gory, his descriptions including new species of *Astraeus*, *Cyrioides* (as *Chrysochroa*), *Melobasis* (as *Buprestis*), *Paracupta* (as *Chrysodema*), *Prospheres* (as *Buprestis*) and *Xyroscelis* (as *Amorphosoma*). His collection is in the Museum Victoria.

Lea, Arthur Mills (1868–1932). Born in Sydney and died in Adelaide, South Australia. Government Entomologist (New South Wales, Western Australia, Tasmania), Consulting Entomologist (South Australia), Lecturer in Forest Entomology at the University of Adelaide, and Entomologist at the South Australian Museum until his death in 1932. Collected extensively in Australia, as well as Lord Howe Island, Fiji, Malay Peninsula, Java, Borneo, Sulawesi and New Caledonia. Published extensively on Coleoptera and pest insects. Described *Habroloma socialis* (as *Trachys*).

Lesueur, Charles Alexandre (1778–1846). French artist, naturalist and explorer. Collected specimens in Australia, South-east Asia and North America. Lesueur travelled with Francois Péron on Baudin's expedition to the South Seas. In 1846 he was appointed Curator of the Musée d'Histoire Naturelle di Havre.

Lucas, Arthur Henry Shakespeare (1853–1936). School master, zoologist and botanist. Born in Stratford-on-Avon, England. Arrived in Australia in 1883 and held teaching posts in Sydney and Melbourne. Studied Australian algae. Edited the *Victorian Naturalist* from 1884 to 1892. Journeyed in Tasmania with H. J. Carter.

Lumholtz, Carl Sofus (1851–1922). Born in Lillehammer, Norway. A naturalist who during the 1880s collected in Queensland for the University of Christiana, Norway. Described *Temognatha alternata* (as *Stigmodera*).

Macleay, William Sharp (1792–1865) and **Macleay**, William John (1820–91). Patrons of science in Australia. William Sharp Macleay was the eldest son of Alexander Macleay, was born in London and moved to Sydney in 1839, studying marine life and collecting insects. He became a trustee of the Australian Museum (1853–62). William John Macleay was the nephew of Alexander Macleay, born in Scotland but moving to Australia where he died at Sydney in 1891. Described taxa resulting from the Macleay Collection includes species of *Polycesta*, *Castiarina*, *Chrysobothris*, *Habroloma*, *Nascioides*, *Neocuris*, *Notographus*, *Paracephala* and *Pseudanilara*; though a number of the original species descriptions have now been transferred to other genera.

Mannerheim, Carl Gustaf von (Graf) (1797–1854). Finnish Coleopterist and governor of the Viipuru province in the Grand Duchy of Finland. Died at Stockholm, Sweden. Described *Selagis spencei* and *Temognatha obscuripennis* (as *Stigmodera*).

Masters, George (1837–1912). Born in Kent, England and died in Sydney. Collected insects at Port Denison, the County of Cumberland, and in 1875 he visited north-eastern Australia and New Guinea, collecting for Sir William Macleay at Gayndah, Queensland. In 1864 he joined the Australian Museum as Curator and Collector, in 1874 took up the position of Curator of the Macleay Collection, and on the transfer of that collection to the University of Sydney became the Curator for Life of the university's Macleay Museum. In 1885 Masters published the first part of what would be the *Catalogue of the Described Coleoptera of Australia*, published in seven parts, the last in 1887 but with two later supplements in 1895, 1896. *Agrilus mastersii*, *Astraeus mastersii*, *Chrysobothris mastersii* and *Polycesta mastersii* were named in his honour.

Obenberger, Jan (1892–1964). Curator of the Entomological Department in the Museum of Prague. Obenberger published numerous papers on the Australian buprestid fauna (e.g. *Agrilus*, *Anilara*, *Anthaxoschema*, *Bubastes*, *Castiarina*, *Chrysobothris*, *Cisseoides*, *Dinocephalia*, *Germarica*, *Melobasis*, *Neocuris*, *Notographus*, *Pseudanilara*, *Selagis*) though many of his names, including genera (e.g. *Briseis*, *Neotorresita*, *Pseudsynechocera*), proved synonyms of earlier named taxa and often at odds with the opinions of André Théry and H. J. Carter, an example of counter determinations being that of Carter (1928a).

Oke, Charles (1888–1959). Born at Beechworth, Victoria. Worked as an entomologist in Melbourne during the 1920s, Curator at the National Museum of Victoria. Particularly interested in Staphylinidae. His collection is in Museum Victoria.

Olliff, Arthur Sidney (1865–95). Born in Hampshire, England, and died in Sydney. As a young boy he was employed by E. W. Janson to set insects and later undertook similar work at the British Museum. In 1884 he left England to commence as Assistant Zoologist (Entomology) at the Australian Museum. In 1890 he was appointed Government Entomologist at the Agricultural Department of New South Wales, filling this position at the time of his death. Olliff published papers on the insect fauna of Lord Howe Island, as well as Coleoptera (e.g. Cerambycidae, Cetoninae, Oedemeridae, Ptinidae, Staphylinidae) and other groups including Lepidoptera. Described buprestids include *Melobasis empyria*, *Nascio chydaea*, *Nascioides multesimus* (as *Nascio*) and *N. mundus* (as *Nascio*). *Nascioides olliffi* was named in his honour.

Parry, Frederick John Sidney (1810–85). Born in England. Served from 1831 to 1835 in the British Army. Entomological interests included Buprestidae, Cetoniinae and Lucanidae. Described *Metaxymorpha grayii* (as *Stigmodera*). *Nascioides parryi* was named in his honour.

Péron, Francois (1775–1810). French naturalist, explorer and historian. Appointed natural historian and anthropologist in Nicolas Baudin's expedition to the South Seas. 100 000 specimens were amassed by Péron and his assistants during the voyage. *Neobuprestis peroni*, *Chrysobothris peroni* and *Selagis peroni*, the latter described from Kangaroo Island, were named in his honour.

Poll, Jacob R. H. Neervoort van de (1862–1924). Dutch entomologist. Amassed a large beetle collection, and published descriptions of *Astraeus*, *Nascio*, *Nascioides*, *Trachys* and *Pseudotaenia*. *Astraeus polli* and the subgenus *Depollus* (of *Astraeus*) were named in his honour.

Rainbow, William Joseph (1856–1919). Born in Yorkshire, England, then moved to New Zealand, and after 10 years there moved to Australia in 1883. Entomologist to the Australian Museum (1895–1919). Mainly interested in Arachnida. Described *Castiarina cydista* (as *Stigmodera*).

Saunders, Edward (1848–1910). Born at Wandsworth, England. Saunders described a large number of Australian Buprestidae including species currently placed in *Astraeus*, *Calotemognatha*, *Castiarina*, *Chalcophorotaenia*, *Melobasis*, *Paratrachys*, and *Pseudotaenia*. Species named in his honour include *Bubastes saundersi* and *Chrysobothris saundersii*.

Skuse, Frederick Askew (1864–96). Arrived in Australia from England in about 1886. In 1890 he was appointed as a scientific assistant in the Entomological Department of the Australian Museum, succeeding

Arthur Olliff. Specialised in Diptera. *Castiarina skusei* is named in his honour.

Sutton, Edmund (1888–1981). Amateur entomologist. Lived at Fletcher, southern Queensland. *Castiarina suttoni* and *Theryaxia suttoni* were named in his honour.

Tepper, Johann Gottlieb Otto (1841–1922). Born in Posen, Germany. Died in Adelaide 1922. In 1883 appointed to the staff of the South Australian Institute Museum (later renamed the South Australian Museum) as a natural history collector. In 1888 he was appointed to the position of entomologist and librarian and also acted as a consulting entomologist to the Department of Agriculture. Published papers on various insects and wrote the *Common Native Insects of South Australia: a Popular Guide to South Australian Entomology* (two parts). *Castiarina tepperi* was named in his honour.

Théry, André (1864–1947). French entomologist. Théry was an agricultural engineer at Rabat, Morocco, described a number of Australian Buprestidae (e.g. *Anilara antiqua*, *Anthaxoschema carteri*, *Bubastes bostrychoides* [as *Neraldus*], *Cyphogastra pistor vulnerata*, *Euryspilus caudatus*, *Mastogenius frenchi*, *Melobasis chrysomelina*, *Temognatha apicenigra* [as *Stigmodera*], *Synechocera tasmanica*) and contributed to H. J. Carter's 1929 checklist of Australian Buprestidae. The genus *Theryaxia* (*T. suttoni*) was named by Carter in his honour.

Thomson, James (1828–97). American by birth but lived in France for most of his life. Wrote papers on Coleoptera, describing Australian Buprestidae (e.g. *Astraeus navarchis*, *Bubastes inconsistans*, *Chrysobothris amplicollis*, *Germarica lilliputana* (as *Aphanisticus*), *Julodimorpha saundersii*, *Microcastalia globithorax* (as *Castalia*), *Synechocera elongata*, *Temognatha coelesta* (as *Stigmodera*) and a number of *Melobasis*. His extensive private collection of Coleoptera was sold to M. René Overthür. Species named in his honour include *Bubastes thomsoni* and *Castiarina thomsoni*.

Thunberg, Carl Pehr (1743–1828). Born in Sweden and was the successor to Linné as Professor of Natural History in Upsala. Named *Belionota prasina* (as *Buprestis*).

Tillyard, Robin John (1881–1937). Highly distinguished scholar, university lecturer and entomologist. Born at Norwich, England and moved to Australia in 1904, there succeeding H. J. Carter at Sydney Grammar School. Tillyard published on almost all insect orders but had a particular interest in dragonflies, lacewings, the evolution of insects and fossil insects. Tillyard discovered *Nascioides costatus* and *N. tillyardi*.

Walker, James John (1851–1939). English naturalist and entomologist, officer in the Royal Navy and collected insects during his various voyages. Later editor of the *Entomologist's Monthly Magazine*. While stationed in Australia he collected with H. J. Carter in the Illawarra and Ourimbah areas. His collections from the Pacific, Australia, New Zealand and the Mediterranean were shared between the Natural History Museum, London, and Oxford University Museum.

Waterhouse, Charles Owen (1843–1917). Entomologist but interested chiefly in Coleoptera. Born in London, he was appointed Junior Assistant in the Entomological Department of the British Museum rising to Assistant Keeper in 1905 and retiring in 1910. His papers dealt with several families including Buprestidae (e.g. *Castiarina atronotata* [as *Stigmodera*], *Chalcotaenia quadriimpressa*, *Nascioides carissimus* (as *Nascio*), *Selagis corusca* [as *Curis*]), Scarabaeidae, Mordellidae and Carabidae.

Westwood, John Obadiah (1805–93). Born in Sheffield, England. Entomologist and first Hope Professor at Oxford, Chair of Invertebrate Zoology. Wrote prolifically on Australian Entomology. Described *Castiarina decipiens* (as *Stigmodera*). *Diphucrania westwoodii* and *Temognatha westwoodii* were named in his honour.

White, Adam (1817–79). Born in Edinburgh, Scotland, serving on the staff of the Zoological Department of the British Museum from 1835 to his retirement in 1863. Described *Castiarina parallela*, *Diadoxus erythrurus*, *Julodimorpha bakewellii* and *Temognatha conspicillata* (all as *Stigmodera*).

Wilson, Francis Erasmus (1888–1960). Entomologist. Born at Beechworth, Victoria. Collected in Queensland, New South Wales, Victoria and South Australia. President of the Field Naturalist Club of Victoria and the Entomological Society of Victoria. His collection is housed in Museum Victoria.

Zeck, Emil Hermann (1891–1963). Born in Sydney. Entomological and biological artist, and entomologist with the New South Wales Department of Agriculture. Collaborated with H. J. Carter, H. W. Eastwood and A. Musgrave on studies of Australian insects. *Castiarina zecki* and *Selagis zecki* were named in his honour.

Appendix 5. Divisions of geological time

(Adapted from Hansen 1991 and Barlow 1981).

Era	Period	Epoch	Elapsed time since beginning of the period or epoch
Cenozoic	Quaternary	Holocene	11 400 years
		Pleistocene	1.8 my*
	Tertiary	Pliocene	5.3 my
		Miocene	23.3 my
		Oligocene	33.9 my
		Eocene	55.8 my
		Palaeocene	65.5 my
Mesozoic	Cretaceous	Late/Upper	99.6 my
		Early/Lower	145.5 my
	Jurassic	Late/Upper	161.2 my
		Middle	175.6 my
		Early/Lower	199.6 my
	Triassic	Late/Upper	228 my
		Middle	245 my
		Early/Lower	251 my
Palaeozoic	Permian		299 my
	Carboniferous	Late/Upper	318 my
		Early/Lower	359.2 my
	Devonian		416 my
	Silurian		443.7 my
	Ordovician		488 my
	Cambrian		542 my

* my = millions of years

Glossary

abiotic: without the aid of biological agents, thus by wind, gravity or water.

actinomorphic: radially symmetrical; describes a flower that can be bisected in a number of planes, with sepals and petals of the same or largely similar shape. cf. **zygomorphic**.

angiosperm: a flowering, seed-bearing, vascular plant characterised by ovules enclosed in ovaries; distinct from gymnosperms.

anthophilous: 'flower loving', flower-frequenting, used in reference to animals (including insects) that visit and usually feed on flowers; but may not be true pollinators

apomorphic: derived from and differing from an ancestral condition.

aposematic: colouration that has a defensive/protective function.

appendiculate: bearing appendages (e.g. claws).

biogeographical region: any geographical area or division characterised by distinctive fauna and or flora

buprestins: 'bitter principles' of a novel, general structure so named by Moore and Brown (1985). Often a volatile oil, alkaloid, sesquiterpene, or some combination thereof. 'Bitter principles' are any one of several hundred compounds having a bitter taste, and not admitting of any chemical classification.

carina (pl. carinae): an elevated ridge or keel.

cf.: compare.

co-evolutionary: describes the interdependent evolution of two or more species having an obvious ecological relationship; usually interactively beneficial.

constancy: temporal response of a pollinator, or flower visitor, to a particular flowering plant when foraging for resources. Unlike **fidelity** the relationship is not fixed, the pollinator potentially switching to other flower species as they become available or more abundant.

convergence: the independent evolution of functional or structural similarity in two or more unrelated or distantly related forms/lineages.

cuticle: superficial non-cellular layer secreted by the epidermis; protects against mechanical injury but chief function is in preventing excessive water loss.

dioecious: of a species separate, having male or female plants.

diurnal: active during daylight, as opposed to at night.

elytra: in Coleoptera, the anterior leathery or chitinous wings, serving as a protective covering to the hind wings.

exine: generally the outer-most structural layer of the pollen grain, secondnd layer if pollenkitt is considered as a layer.

extralimital: outside or extra to the normal or focal geographical distribution of a species.

fidelity: a fixed relationship (though often used synonymously with **constancy**); the degree of restriction to an anthophilous insect (or other flower-visiting animal) to a particular plant species or higher taxon. Fidelity to plant hosts is clearly indicated by many buprestid taxa with respect to their larval stages.

florivorous: flower-associated, feeding on floral resources.

frontoclypeus: frons and clypeus (loosely the 'foreface'), frequently delineated by a suture.

frontovertex: combined front and top of the head; between the eyes.

geitonomous: relating to geitonogamy, the transfer of pollen grains from the anther to the stigma of another flower of the same plant; a form of self-pollination.

genal: relating to the cheek (gena), the part of the head on each side below the eyes.

Gondwana: the supercontinent variously through geological time including Australia, Africa, Madagascar, India, New Zealand, New Caledonia, South America etc.

gymnosperm: seed-bearing vascular plant characterised by the non-enclosure of its ovule by a structure resembling the ovary formed by carpels of an angiosperm, but exposed except for one or two integuments on the surface of the sporophylle. Lacks true flowers.

maculation: a pattern that is spotted or marked with figures that differ in colour or shape to the ground colour.

maxilla (pl. maxillae): the second pair of jaws in a mandibulate, mandible-bearing, insect.

mesic: used in relation to plants that inhabit and are adapted to moist environments, but here restricted to rainforest and wet sclerophyll associated species.

mesothorax: the second or middle thoracic ring which bears the middle legs and the anterior legs.

metacoxal: relating to the basal segment of the hind leg by means of which it is articulated to the body.

microhymenoptera: small wasps several millimetres or less in size; not readily visible to the naked eye.

monoecious: having separate female and male flowers on the same plant.

mya: million years ago.

myrmecophily: pollination by ants.

ovipositor: the tubular or valved structure by means of which the eggs are placed.

parasitoid: an organism with a mode of life intermediate between parasitism and predation, as in Hymenoptera in which the larva feeds within the living body of another organism, eventually causing the death of the host.

pharate: where an insect has undertaken complete metamorphosis to the adult stage but is still within the pupa.

phenological event: a temporal recurrent natural phenomenon; related to weather and climate.

phyllophagous: feeding on leaves.

phytophagous: feeding on plants or plant material; herbivorous.

pollenkitt: an oily, sometimes pigmented, coating of pollen grains, but not always present; sometimes considered the outer-most layer of the pollen grain.

poricidal anthers: anthers with only a small opening at the end.

procrypsis: pertaining to colouration and behaviour that affords protection against predators.

prognathous: of insects, often Coleoptera, in which the head is horizontal with the 'muzzle-like' jaws directed downward.

pronotum: the upper or dorsal surface of the prothorax.

prosternum: the 'fore-breast', the sclerite between the forelegs.

prothorax: the first thoracic segment or 'thorax'; bears the anterior legs but no wings.

pulvilli (sing. pulvillus): pad-like structure between the tarsal claws, or on undersides of tarsal joints.

punctuation: markings of small pits.

pygidium: the dorsal surface of the last segment of the abdomen.

saltatorial: adapted for a leaping or bounding motion.

scrobes: grooves formed for the reception or concealment of an appendage (antennae).

serrate: with notched edges like the teeth of a saw.

setae: commonly known as 'hairs'; slender hair-like appendages.

tarsus (pl. tarsi): the jointed appendage attached to the apex of the tibia.

trap-lining: movements by a pollinator between flowers (the resource being 'trapped') of the same species, which may be widely spaced. The association of the pollinator/visitor is not necessarily fixed to a particular flowering taxon, but may differ with respect to temporal availability and spatial distribution of available flowering plants.

unisexual: of a flower that is either male or female. cf. **hermaphrodite**, **bisexual**.

xenogamous: exhibiting cross-fertilisation between flowers on different plants.

xeric-adapted: tolerating or adapted to dry conditions.

zygomorphic: bilaterally symmetrical; pertaining to a flower in which the petals and sepals differ in their shape, size and/or number such the flower can only be bisected in a single plane; cf. **actinomorphic**.

Bibliography

Adam P (1987) *New South Wales Rainforests: The Nomination for the World Heritage List*. National Parks and Wildlife Service of New South Wales, Sydney.

Adam P (1992) *Australian Rainforests*. Clarendon Press, Oxford.

Adam P, Williams G (2001) Dioecy, self-compatibility and vegetative reproduction in Australian subtropical rainforest trees and shrubs. *Cunninhamia* **7**, 89–100.

Alexeev AV (2000) On Mesozoic buprestids (Coleoptera: Buprestidae) from Russia, Kazakhastan, and Mongolia. *Palaeontological Journal* **34**, S323–S326.

Alexeev AV (2009) New jewel beetles (Coleoptera: Buprestidae) from the Cretaceous of Russia, Kazakhastan, and Mongolia. *Palaeontological Journal* **43**, 277–281. doi:10.1134/S0031030109030058

Archibold OW (1995) *Ecology of World Vegetation*. Chapman and Hall, London.

Arroyo-Correa B, Burkle LA, Emer C (2019) Alien plants and flower visitors disrupt the seasonal dynamics of mutualistic networks. *Austral Ecology* **108**, 1475–1486. doi:10.1111/1365-2745.13332

Barker S (1975) Revision of the genus *Astraeus* Laporte and Gory (Coleoptera: Buprestidae). *Transactions of the Royal Society of South Australia* **99**, 105–142.

Barker S (1977) *Astraeus* (Coleoptera: Buprestidae): a description of three new species and new locality records. *Transactions of the Royal Society of South Australia* **101**, 11–14.

Barker S (1979) New species and a catalogue of *Stigmodera* (*Castiarina*) (Coleoptera: Buprestidae). *Transactions of the Royal Society of South Australia* **103**, 1–23.

Barker S (1980) New species and new synonyms of *Stigmodera* (Castiarina) (Coleoptera: Buprestidae). *Transactions of the Royal Society of South Australia* **104**, 1–7.

Barker S (1983) New synonyms and new species of *Stigmodera* (*Castiarina*) (Coleopetra: Buprestidae). *Transactions of the Royal Society of South Australia* **107**, 139–169.

Barker S (1986) *Stigmodera* (*Castiarina*) (Coleoptera: Buprestidae): taxonomy, new species and a checklist. *Transactions of the Royal Society of South Australia* **110**, 1–36.

Barker S (1989) Contributions to the taxonomy of Australian Buprestidae (Coleoptera): new species of *Astraeus* and *Stigmodera* (*Castiarina*) and a key to *Astraeus* (s.s.). *Transactions of the Royal Society of South Australia* **113**, 185–194.

Barker S (1990) Two replacement names in *Castiarina* (Buprestidae: Coleoptera). *Transactions of the Royal Society of South Australia* **114**, 105.

Barker S (1993a) Seventeen new species of Australian Buprestidae (Insecta: Coleoptera) and a new host for *Castiarina uptoni* (Barker). *Transactions of the Royal Society of South Australia* **117**, 15–26.

Barker S (1993b) A new Australian species of *Calodema* (Coleoptera: Buprestidae). *Transactions of the Royal Society of South Australia* **117**, 191–192.

Barker S (1995) Eight new species of Australian Buprestidae (Insecta: Coleoptera). *Transactions of the Royal Society of South Australia* **119**, 149–156.

Barker S (1996) Seventeen new species of *Castiarina* (Coleoptera: Buprestidae). *Transactions of the Royal Society of South Australia* **120**, 41–59.

Barker S (1999) Designation of a lectotype and descriptions of four new species of Australian Buprestidae (Coleoptera). *Records of the South Australian Museum* **32**, 45–49.

Barker S (2001) Descriptions of twenty one new species of *Cisseis* (sensu stricto) Gory & Laporte 1839 (Coleoptera: Buprestidae: Agrilinae). *Transactions of the Royal Society of South Australia* **125**, 97–113.

Barker S (2002) A checklist of *Cisseis* (*sensu stricto*) Gory & Laporte, 1839 (Coleoptera: Buprestidae: Agrilinae). *Records of the South Australian Museum* **35**, 85–90.

Barker S (2006a) *Castiarina: Australia's Richest Jewel Beetle Genus*. Australian Biological Resources Study, Canberra.

Barker S (2006b) Twenty five new species of *Cisseis* (*sensu stricto*) and two new synonyms (Coleoptera: Buprestidae: Agrilinae). *Transactions of the Royal Society of South Australia* **131**, 257–284. doi:10.1080/3721426.2006.10887065

Barker S (2006c) Five new species of Australian Buprestidae (Coleoptera). *Transactions of the Royal Society of South Australia* **131**, 285–296. doi:10.1080/3721426.2006.10887066

Barker S, Bellamy CL (2001) *Stanwatkinsius*, a new genus of Australian jewel beetles (Coleoptera: Buprestidae: Agrilinae) with a key to known species. *Transactions of the Royal Society of South Australia* **125**, 1–14.

Barker S, Inns, R (1976) Predation on *Stigmodera* (*Themognatha*) *tibialis* by a fly. *The Western Australian Naturalist* **13**, 147–148.

Barlow BA (1981) The Australian flora: its origin and evolution. In *Flora of Australia*. Vol. 1. Introduction. Australian Government Publishing Service, Canberra.

Barreda VD, Palazzesi L, Tellaria MC, Olivero EB, Raine JI, *et al.* (2015) Early evolution of the angiosperm clade Asteraceae in the Cretaceous of Antarctica. *Proceedings of the National Academy of Sciences of the United States of America* **112**, 10989–10994. doi:10.1073/pnas.1423653112

Barringer L (2016) First record of *Mastogenius crenulatus* (Knull) (Coleoptera: Buprestidae) in Massachusetts. *Insecta Mundi* **0520**, 1–2.

Beadle NCW, Evans OD, Carolin RC (1972) *Flora of the Sydney Region*. AH and AW Reed, Sydney.

Beard JS (1990) *Plant Life of Western Australia*. Kangaroo Press, Kenthurst.

Bellamy CL (1986) The higher classification of Australian Buprestidae, with the description of a new genus and species (Coleoptera). *Australian Journal of Zoology* **34**, 583–600. doi:10.1071/ZO9860583

Bellamy CL (1987a) Two new species of *Anthaxomorphus* Deyrolle from southern Africa with a new synonym and a world checklist (Coleoptera: Buprestidae: Trachyderinae). *Coleopterists Bulletin* **41**, 287–295.

Bellamy CL (1987b) A revision of the genus *Synechocera* Deyrolle (Coleoptera: Buprestidae: Agrilinae). *Invertebrate Taxonomy* **1**, 17–34. doi:10.1071/IT9870017

Bellamy CL (1988) The classification and phylogeny of the Australian Coroebini, Bedel, with a revision of the genera *Paracephala*, *Meliboeithon* and *Dinocephalia* (Coleoptera: Buprestidae: Agrilinae). *Invertebrate Taxonomy* **2**, 413–453. doi:10.1071/IT9880413

Bellamy CL (1990) Studies in the Mastogeniinae (Coleoptera: Buprestidae) III. New species, combinations and a world category. *Giornale Italiano di Entomologia* **5**, 109–128.

Bellamy CL (1991) Further review of the genus *Maoraxia* Obenberger (Coleoptera: Buprestidae). *Invertebrate Taxonomy* **5**, 457–468. doi:10.1071/IT9910457

Bellamy CL (1994) *Balthasarella melandryoides* Obenberger: a relict buprestid becomes less enigmatic (Coleoptera: Buprestidae). *Coleopterists Bulletin* **48**, 300.

Bellamy CL (1995) Buprestidae (Coleoptera) from amber deposits: a brief review and family switch. *The Coleopterists Bulletin* **49**, 175–177.

Bellamy CL (1997) Phylogenetic relationships of *Xyroscelis* (Coleoptera: Buprestidae). *Invertebrate Taxonomy* **11**, 569–574. doi:10.1071/IT94026

Bellamy CL (1999) A new species of Buprestidae (Coleoptera) from Dominican amber. *Coleopterists Bulletin* **53**, 321–323.

Bellamy CL (2002) *Coleoptera: Buprestoidea. In Zoological Catalogue of Australia* Vol. 29.5. (Ed. WWK Houston). CSIRO Publishing, Melbourne.

Bellamy CL (2003) *An Illustrated Summary of the Higher Classification of the Superfamily Buprestoidea (Coleoptera). Folia Heyrovskyana Supplementum 10.*

Bellamy CL (2005) Clarification of synonymy in three species of *Temognatha* Solier, 1833 (Coleoptera: Buprestidae). *The Pan-Pacific Entomologist* **81**, 99–100.

Bellamy CL (2006) Studies on the Australian Chalcophorini: a new genus for *Chalcophora subfasciata* Carter, 1916 and a review of the *Pseudotaenia* Kerremans 1893 generic-group (Coleoptera: Buprestidae). *Zootaxa* **1206**, 23–46.

Bellamy CL (2008a) *A World Catalogue and Bibliography of the Jewel Beetles (Coleoptera: Buprestoidea), Volume 1: Introduction; Fossil Taxa; Schizopodidae; Buprestidae: Julodinae – Chrysochroinae: Poecilonotini*. pp. 1–625. Pensòft (Series Faunistica No. 76), Sofia.

Bellamy CL (2008b) *A World Catalogue and Bibliography of the Jewel Beetles (Coleoptera: Buprestidoidea), Volume 2: Chyrsochroinae: Sphenopterini through Buprestinae: Stigmoderini*. pp. 626–1259. Pensòft (Series Faunistica No. 77), Sofia.

Bellamy CL (2008c) *A World Catalogue and Bibliography of the Jewel Beetles (Coleoptera: Buprestidoidea), Volume 3: Buprestinae: Pterobothrini through Agrilinae: Rhaeboscelina*. pp. 1260–1931. Pensòft (Series Faunistica No. 78), Sofia.

Bellamy CL (2008d) *A World Catalogue and Bibliography of the Jewel Beetles (Coleoptera: Buprestidoidea), Volume 4: Agrilinae: Agrilina through Trachyini*. pp. 1932–2684. Pensòft (Series Faunistica No. 79), Sofia.

Bellamy CL (2009) *A World Catalogue and Bibliography of the Jewel Beetles (Coleoptera: Buprestidoidea), Volume 5: Appendices, Bibliography, Indices*. pp. 2685–3264. (Series Faunistica No. 80), Sofia.

Bellamy CL (2013) *Aaaaba* Bellamy, a second replacement name for *Alcinous* Deyrolle, 1864 (Coleoptera: Buprestidae). *The Coleopterists Bulletin* **67**, 32. doi:10.1649/072.067.0107

Bellamy CL, Nylander U (2007) New genus-group synonymy in Stigmoderini (Coleoptera: Buprestidae). *The Coleopterist's Bulletin* **61**, 423–427.

Bellamy CL, Peterson M (2000a) Contributions to the taxonomy of Australian Buprestidae (Coleoptera) Part I. New synonymy, combinations, and names. *Folia Heyrovskyana* **8**, 73–100.

Bellamy CL, Peterson M (2000b) Contributions to the taxonomy of Australian Buprestidae (Coleoptera) Part II. Three new species and a new coraebine generic member of the Australian fauna. *Folia Heyrovskyana* **8**, 101–108.

Bellamy CL, Weir TA (2008) The reinstatement of *Julodimorpha saundersii* Thomson 1879 (Coleoptera: Buprestidae) as a valid species. *Zootaxa* **1751**, 46–54. doi:10.11646/zootaxa.1751.1.4

Bellamy CL, Williams GA (1985) A revision of *Maoraxia* with a new synonym in *Acmaeodera* (Coleoptera: Buprestidae). *International Journal of Entomology* **27**, 147–161.

Bellamy CL, Williams GA (1995) The first Australian *Paratrachys* (Coleoptera: Buprestidae), with comments on the Higher Classification of the genus. *Invertebrate Taxonomy* **9**, 1265–1276.

Bellamy CL, Williams GA, Hasenpusch J, Sundholm A (2013) A summary of the published data on host plants and morphogy of immature stages of Australian jewel beetles (Coleoptera: Buprestidae), with additional new records. *Insecta Mundi* **0293**, 1–172.

Bernhardt P (2000) Convergent evolution and adaptive radiation of beetle-pollinated angiosperms. *Plant Systematics and Evolution* **222**, 293–320. doi:10.1007/BF00984108

Biffin E, Lucas EJ, Craven LA, Ribeiroda Costa I, Harrington MG, *et al.* (2010) Evolution of exceptional species richness among lineages of fleshy-fruited Myrtaceae. *Annals of Botany* **106**, 79–93. doi:10.1093/aob/mcq088

Bílý S (1974) Zur biologie einheimischer Kaferfamilien 13. Buprestidae. *Entomologische Berichten* **2**, 67–78.

Bílý S (1986) Descriptions of adult larvae of *Thrincopyge alacris* LeConte and *Aphanisticus cochinae seminulum* Obenberger (Coleoptera, Buprestidae). Entomological papers presented to Yoshihiko Kurosawa on the occasional of his retirement March 20, 1986. Coleopterists Association of Japan, Tokyo.

Bílý S (1989) Descriptions of the last instar larvae of *Polycesta porcata* and *Paratrachys hederae hederae* (Coleoptera, Buprestidae). *Acta Entomologica Bohemoslovaca* **86**, 61–66.

Bílý S (1999) Larvae of buprestid beetles (Coleoptera: Buprestidae) of Central Europe. *Acta Entomologica Musei Nationalis Pragae. Supplementum* **9**, 1–107.

Bílý S (2000) A new concept of Anthaxiini (Coleoptera: Buprestidae). *Folia Heyrovskyana* **8**, 109–114.

Bílý S, Fikacek M, Šípek P (2008) First record of myrmecophily in buprestid beetles: immature stage of *Habroloma myrmecophila* sp. nov. (Coleoptera: Buprestidae) associated with *Oecophylla* ants (Hymenoptera: Formicidae). *Insect Systematics and Evolution* **39**, 121–131. doi:10.1163/187631208788784084

Bílý S, Hanlon M (2020) A revision of the genus *Bubastes* Laporte and Gory, 1836 (Coleoptera: Buprestidae). *Zootaxa* **4832**, 1–75. doi:10.11646/zootaxa.4832.1.1

Bílý S, Powell M (2017) A new species of the genus *Neobubastes* Blackburn, 1892 from Western Australia (Coleoptera: Buprestidae: Buprestinae, Pterobothrini). *Zootaxa* **4329**, 86–90. doi:10.11646/zootaxa.4329.1.5

Bílý S, Volkovitsh MG (1996) Revision, reclassification and larval morphology of the genus *Paratassa* (Coleoptera: Buprestidae: Paratassini tribus n.). *Acta Societatis Zoologicae Bohemoslovenicae* **60**, 325–346.

Bílý S, Volkovitsh MG (2003) Larvae of Australian Buprestidae (Coleoptera). Part 1. Genera *Austrophorella* and *Pseudotaenia*. *Acta Societatis Zoologicae Bohemoslovenicae* **67**, 99–144.

Bílý S, Volkovitsh MG (2005) Larvae of Australian Buprestidae (Coleoptera). Part 1. Genera *Maoraxia* and *Anthaxoschema*. *Folia Heyrovskyana Serie A* **13**, 7–26.

Bílý S, Volkovitsh MG (2007) Descriptions of some buprestid larvae from Chile. *Folia Heyrovskyana* **15**, 53–79.

Bílý S, Volkovitsh MG (2013) Larvae of Australian Buprestidae (Coleoptera). Part 4. Genus *Julodimorphus*. *Zootaxa* **3637**, 341–354. doi:10.11646/zootaxa.3637.3.6

Blackburn T (1892) Further notes on Australian Coleoptera, with descriptions of new genera and species, xii. *Transactions of the Royal Society of South Australia* **15**, 207–261.

Briese DT (1991) Current status of *Agrilus hyperici* (Coleoptera: Buprestidae) released in Australia in 1940 for the control of St. John's wort: lessons for insect introductions. *Biocontrol Science and Technology* **1**, 207–215. doi:10.1080/09583159109355200

Britton EB (1970) Coleoptera. In *Insects of Australia*. (Ed. Naumann I). pp. 495–621. Melbourne University Press, Melbourne.

Brooks JG (1948) Some North Queensland Coleoptera and their food plants. *The North Queensland Naturalist* **15** 26–29.

Brown WV, Jones AJ, Lacey MY, Moore BP (1985) Chemistry of buprestins A and B. Bitter principles of jewel beetles (Coleoptera: Buprestidae). *Australian Journal of Chemistry* **38**, 197–206. doi:10.1071/CH9850197

Brues CT, Melander AL, Carpenter FM (1954) Classification of insects. Bulletin of the Museum of Comparative Zoology, Harvard. **108**, 1–917.

Burns GG, Burns AJ (1992) The distribution of Victorian jewel beetles (Coleoptera: Buprestidae) – an ENTRECS project. *Occasional Papers from the Museum of Victoria* **5**, 1–53.

Cai C, Ślipiński A, Huang D (2015) First false jewel beetle (Coleoptera: Schizopodidae) from the Lower Cretaceous of China. *Cretaceous Research* **52**, 490–494. doi:10.1016/j.cretres.2014.03.028

Cameron DG (1987) Temperate rainforests of East Gippsland. Chapter 4. In *The Rainforest Legacy*, Vol. 1 (Eds GL Werren, AP Kershaw). pp. 33–46. Australian Governmant Publishing Service, Canberra.

Carnaby K (1987) *Jewel Beetles of Western Australia*. Privately printed K. and E. Carnaby, Wilga.

Carter HJ (1906) Notes on the genus *Cardiothorax*: with descriptions of new species of Australian Coleoptera. *Proceedings of the Linnean Society of New South Wales* **31**, 236–260.

Carter HJ (1908) Revision of the genus *Seirotrana*, together with descriptions of new species of other Australian Coleoptera. *Proceedings of the Linnean Society of New South Wales* **33**, 392–422.

Carter HJ (1915) Descriptions of six new species of Buprestidae. *Proceedings of the Linnean Society of New South Wales* **40**, 76–82.

Carter HJ (1916) Revision of the genus *Stigmodera*, and description of some new species of Buprestidae (Order Coleoptera). *Transactions and Proceedings of the Royal Society of South Australia* **40**, 78–144.

Carter HJ (1920) Notes on some Australian Tenebrionidae, with descriptions of new species; – also of a new genus and species of Buprestidae. *Proceedings of the Linnean Society of New South Wales* **45**, 222–249.

Carter HJ (1923a) Revision of the genera *Ethon*, *Cisseis* and their allies (Buprestidae). *Proceedings of the Linnaean Society of New South Wales* **48**, 159–176.

Carter HJ (1923b) A revision of the Australian species of the genus *Melobasis* (Fam. Buprestidae, Order Coleoptera), with notes on allied genera. *Transactions of the Entomological Society of London* **1923**, 64–104.

Carter HJ (1924) Australian Coleoptera – notes and new species No. iii. *Proceedings of the Linnean Society of New South Wales* **49**, 19–45.

Carter HJ (1925) Revision of the Australian species of *Chrysobothris* (Fam. Buprestidae), together with notes and descriptions of new species of Coleoptera. *Proceedings of the Linnaean Society of New South Wales* **50**, 225–244.

Carter HJ (1926) Revision of the Australasian species of *Anilara* (Fam. Buprestidae) and *Helmis* (Fam. Dryopidae), with notes, and descriptions of other Australian Coleoptera. *Proceedings of the Linnean Society of New South Wales* **51**, 50–71.

Carter HJ (1927) Australian Coleoptera: notes and new species. No. 5. *Proceedings of the Linnean Society of New South Wales* **52**, 222–234.

Carter HJ (1928a) Revision of the Australian species of the genera *Curis*, *Neocuris* and *Trachys*, together with notes and descriptions of new species of other Coleoptera. *Proceedings of the Linnean Society of New South Wales* **53**, 270–290.

Carter HJ (1928b) Revision of *Hesthesis* (Fam. Cerambycidae) together with description of a new genus and species of Buprestidae. *Proceedings of the Linnean Society of New South Wales* **53**, 544–550.

Carter HJ (1929) A check list of the Australian Buprestidae. *Australian Zoologist* **5**, 265–304.

Carter HJ (1931) Notes on the genus *Stigmodera* (Family Buprestidae) together with descriptions of new species of and a retabulation of the subgenus *Castiarina*. *Australian Zoologist* **6**, 337–367.

Carter HJ (1932) New Guinea and Australian Coleoptera. Notes and new species. No. 2. *Proceedings of the Linnean Society of New South Wales* **57**, 101–115.

Carter HJ (1933) *Gulliver in the Bush*. Angus and Robertson, Sydney.

Chanderbali AS, Van der Werff H, Renner SS (2001) Phylogeny and historical biogeography of Lauraceae: evidence from the chloroplast and nuclear genomes. *Annals of the Missouri Botanical Garden* **88**, 104–134. doi:10.2307/2666133

Chang V, Otto AK (1987) A new insect pest of sugarcane found in Hawaii. *Annual Report of the Hawaiian Sugar Planters' Association* **1984**, 36–37.

Christophel DC, Lys SD (1986) Mummified leaves of two new species of Myrtaceae from the Eocene of Victoria, Australia. *Australian Journal of Botany* **34**, 649–662. doi:10.1071/BT9860649

Cobos A (1957) Un genero y dos neuvas especies de Buprestidae de Nueva Guinea. *The Coleopterists Bulletin* **10**, 91–96.

Cobos A (1979) Revision de la subfamilia Trachyinae a niveles suraespecificos (Coleoptera: Buprestidae). *Acta Entomologica Bohemoslovaca* **76**, 414–430.

Cobos A (1980) Ensayo sobrelos géneros de la subfamilia Polycestinae (Coleoptera, Buprestidae) Part 1. *Eos* **54**, 15–94.

Costin, AB, Gray, M, Totterdell, CJ, Wimbush, DJ (2000) *Kosciuszko Alpine Flora*. 2nd edn. CSIRO, East Melbourne and William Collins Pty. Ltd., Sydney.

Cowie D (2001) *Jewel Beetles of Tasmania*. Tasmanian Field Naturalists Club, Hobart.

Cranston PS, Gullan PJ, Taylor RW (1991) Principles and Practice of Systematics. Chapter 4. In *The Insects of Australia* (Ed. ID Naumann). pp. 109–124. Melbourne University Press, Melbourne.

Crisp MD, Cook LG (2011) Cenozoic extinctions account for the low diversity of extant gymnosperms compared with angiosperms. *New Phytologist* **192**, 997–1009. doi:10.1111/j.1469-8137.2011.03862.x

Crous KY (2019) Plant responses to climate warming: physiological adjustments and implications for plant functioning in a future, warmer world. *American Journal of Botany* **106**, 1049–1051. doi:10.1002/ajb2.1329

Crowson RA (1981) *Biology of the Coleoptera*. Academic Press, London.

Curletti G (2001) The genus *Agrilus* in Australia. *Jewel Beetles* **9**, 1–45.

Curletti G (2002) The genus *Agrilus* in Australia. *Jewel Beetles* **11**, 1–4.

Curletti G (2010) A new species of *Agrilus* Curtis (Coleoptera: Buprestidae) from eastern Australia. *Australian Entomologist* **36**, 167–169.

Curran TJ, Clarke PJ, Bruhl JJ (2008) A broad typology of dry rainforests on the western slopes of New South Wales. *Cunninghamia* **10**, 381–405.

Doherty MD, Wright G, McDougall KL (2015) The flora of Kosciuszko National Park, New South Wales: summary and overview. *Cunninghamia* **15**, 13–68. doi:10.7751/cunninghamia.2015.15.002

Douglas AM (1980) *Our Dying Fauna*. Imperial Publishing Company, Perth.

Duke NC (2006) *Australia's Mangroves: The Authoritative Guide to Australia's Mangrove Plants*. University of Queensland and Norman C. Duke, Brisbane.

Dumbleton LJ (1932) Early stages of New Zealand Buprestidae. *Stylops* **1**, 41–48.

Dunstan B (1923) Mesozoic insects of Queensland. Part 1. Introduction and Coleoptera. *Queensland Geological Survey Publication* No. 273.

Enright NJ, Hill RS, Veblen TT (1995) The southern conifers – an introduction. In *Ecology of Southern Conifers*. (Eds NJ Enright, RS Hill). pp. 1–9. Melbourne University Press, Carlton.

Etheridge R, Olliff AS (1890) The Mesozoic and Tertiary Insects of New South Wales. *Memoirs of the Geological Survey of New South Wales* **7**, 1–12.

Evans AM (2017) Molecular phylogeny of the wood-boring beetle family Buprestidae and insights into the evolution of host use. Dissertation in partial fulfilment of the requirements for the degree of Doctor of Philosophy. Harvard University, Cambridge, Massachusetts.

Evans AM, McKenna DD, Bellamy CL, Farrell BD (2015) Large-scale molecular phylogeny of metallic wood-boring beetles (Coleoptera: Buprestidae) provides new insights into relationships and habit. *Systematic Entomology* **40**, 385–400.

Faegri K, Iversen J, Kaland PM, Krzywinski K (1992) *Textbook of Pollen Analysis*. John Wiley and Sons, Chichester.

Fearn S (2016) Tasmania's forgotten jewel: distribution and ecological notes on the jewel beetle *Castiarina bremei* Hope 1845 (Buprestidae: Buprestinae). *The Tasmanian Naturalist* **138**, 68–71.

Ferrière C (1947) A chalcidoid egg-parasite of an Australian buprestid. *Bulletin of Entomological Research* **37**, 629–631. doi:10.1017/S0007485300030121

Fowler WW (1912) *British Coleoptera*. 6 Vols. (1887–1913). *Coleoptera, general introduction and Cicindellidae and*

Paussidae. Fauna of British India. Taylor and Francis, London.

Frank D, Sekerka L (2020) Studies on the genus *Chrysodema* (Coleoptera: Buprestidae: Chrysochroinae). Part 1. *Zootaxa* **4720**, 001–062.

French C (1900) *A Handbook of the Destructive Insects of Victoria, with Notes on the Methods to be Adopted to Check and Extirpate them.* Part III. Government Printer, Melbourne.

Friis EM, Crane PR, Pedersen KR (2011) *Early Flowers and Angiosperm Evolution.* Cambridge University Press, Cambridge.

Gardner JA (1989) Revision of the genera of the tribe Stigmoderini (Coleoptera: Buprestidae) with a discussion of phylogenetic relationships. *Invertebrate Taxonomy* **3**, 291–361.

George A (2008) You don't have to call *Dryandra Banksia. Australian Plants online.* www.anpsa.org.au/APOL2008/sep08-2.html

Gomez JM, Perfectti F, Klingenberg CP (2014) The role of pollinator diversity in the evolution of corolla-shape integration in a pollinator-generalist plant clade. *Philosophical Transactions of the Royal Society B* **369**, 2649 doi:10-1098/rstb.2013-0257

Gory HL, Laporte FL (1840) *Histoire naturelle et iconographie des insectes coléoptéres II.* P. Dumenil, Paris.

Gottsberger G (2015) Generalist and specialist pollination in basal angiosperms (ANITA grade, basal monocots, magnoliids, Chloranthaceae and Ceratophyllaceae): what we know now. *Plant Diversity and Evolution* **131**, 263–362. doi:10.1127/pde/2015/0131-0085

Goudie JC (1920) Notes on Coleoptera of north-western Victoria. *Victorian Naturalist* **37**, 28–34.

Greenwood DR (1987) Early Tertiary Podocarpaceae: megafossils from Eocene Anglesea locality, Victoria, Australia. *Australian Journal of Botany* **35**, 111–133.

Griffiths JR (1974) Revised continental fit of Australia and Antarctica. *Nature* **249**, 336–338.

Grimaldi D, Engel MS (2005) *Evolution of the Insects.* Cambridge University Press, New York.

Grove SJ, Yaxley B (2004) A species of jewel beetle (Coleoptera: Buprestidae) new to Tasmania. *The Tasmanian Naturalist* **126**, 29–30.

Gwynne DT, Rentz DCF (1983) Beetles on the bottle: male buprestids mistake stubbies for females (Coleoptera). *Journal of the Australian Entomological Society* **22**, 79–80.

Hadlington P, Gardner MJ (1959) *Diadoxus erythrurus* (White) (Coleoptera-Buprestidae), attack of fire-damaged *Callitris* spp. *Proceedings of the Linnaean Society of New South Wales* **84**, 325–333.

Hangay G, Zborowski P (2010). *A Guide to the Beetles of Australia.* CSIRO Publishing, Collingwood.

Harden GJ (1990) *Flora of New South Wales.* Vol. 1. New South Wales University Press, Kensington.

Harden GJ (1991) *Flora of New South Wales.* Vol. 2. New South Wales University Press, Kensington.

Harden GJ (1992) *Flora of New South Wales.* Vol. 3. New South Wales University Press, Kensington.

Harden G, McDonald WJF, Williams J (2007) *Rainforest Climbing Plants. A Field Guide to Their Identification.* Gwen Harden Publishing, Nambucca Heads.

Hansen WR (Ed.) (1991) Suggestions to authors of the reports of the United States Geological Survey. 7th edn (STA7). U. S. Geological Survey, Reston, Va.

Hawkeswood TJ (1978) Observations on some Buprestidae (Coleoptera) from the Blue Mountains, N.S.W. *Australian Zoologist* **19**, 257–275.

Hawkeswood TJ (1980a) A spider feeding on a jewel beetle. *The Western Australian Naturalist* **14**, 236.

Hawkeswood TJ (1980b) Jewel beetles as pollinators of *Melaleuca pauperiflora* F. Muell. Between Eucla (W.A.) and Koonalda (S.A.). *The Western Australian Naturalist* **14**, 238–239.

Hawkeswood TJ (1984) New larval host records for two Australian jewel beetles (Coleoptera: Buprestidae). *Coleopterists Bulletin* **39**, 1985.

Hawkeswood TJ (1986a) A new species of *Chrysobothris* (Coleoptera, Buprestidae, Chrysobothrinae) from Townsville, northern Queensland, Australia. *Giornale Italiano di Entomologia* **3**, 167–172.

Hawkeswood TJ (1986b) New larval host records for eight Australian jewel beetles (Coleoptera, Buprestidae). *Giornale Italiano di Entomologia* **3**, 173–177.

Hawkeswood TJ (1987a) Notes on Coleoptera from *Baeckea stenophylla* F. Muell. (Myrtaceae) in New South Wales, Australia. *Giornale Italiano di Entomologia* **3**, 285–290.

Hawkeswood TJ (1987b) *Beetles of Australia.* Angus & Robertson Publishers, Sydney

Hawkeswood TJ (1988) A review of larval host records for twelve Australian jewel beetles (Coleoptera: Buprestidae). *Giornale Italiano di Entomologia* **4**, 81–88.

Hawkeswood TJ (1990) Insect pollination of *Bursaria spinosa* (Pittosporaceae) in the Armidale area, New South Wales, Australia. *Giornale Italiano di Entomologia* **5**, 67–87.

Hawkeswood TJ (1995) A new species of the genus *Chrysobothris* Eschscholtz (Coleoptera: Buprestidae) from Australia. *Giornale Italiano di Entomologia* **7**, 345–349.

Hawkeswood TJ (2002) A review of the biology and host plants of Australian Buprestidae (Coleoptera) known to breed in Eucalyptus species (Myrtaceae). *Journal of the Entomological Research Society* **4**, 31–58.

Hawkeswood TJ (2003) Some notes on the biology and a new host plant of *Cyphogastra bruijni* Lansberge (Coleoptera: Buprestidae) in Papua New Guinea. *Spilopyra* **7**, 1–4.

Hawkeswood TJ (2005) Review of the biology and host plants of the Australian jewel beetle *Julodimorpha bakewelli* (White, 1859) (Coleoptera: Buprestidae). *Calodema* **3**, 3–5.

Hawkeswood TJ (2006a) Review of the biology of *Prospheres aurantiopictus* (Laporte and Gory, 1837) (Coleoptera: Buprestidae). *Calodema* **5**, 3–4.

Hawkeswood TJ (2006b) Review of the biology of the Australian genus *Meliboeithon* Obenberger, 1920 (Coleoptera: Buprestidae). *Calodema* **8**, 3–4.

Hawkeswood TJ (2007a) Review of the biology of the genus *Merimna* Saunders, 1868 (Coleoptera: Buprestidae). *Calodema* **9**, 12–13.

Hawkeswood TJ (2007b) Review of the biology of two Australian species of the genus *Habroloma* Thomson, 1864 (Coleoptera: Buprestidae). *Calodema Supplementary Paper* **4**, 1–2.

Hawkeswood TJ (2007c) Review of the biology of the genus *Anilara* Saunders, 1868 (Coleoptera: Buprestidae). *Calodema Supplementary Paper* **5**, 1–5.

Hawkeswood TJ (2007d) Review of the biology of the genus *Pseudanilara* Théry, 1911 (Coleoptera: Buprestidae). *Calodema* **9**, 9-11.

Hawkeswood TJ (2007e) Review of the biology of the genus *Theryaxia* Carter, 1928 (Coleoptera: Buprestidae). *Calodema* **9**, 14.

Hawkeswood TJ (2007f) Review of the biology of the genus *Torresita* Gemminger and Harold, 1869 (Coleoptera: Buprestidae). *Calodema Supplementary Paper* **21**, 1–2.

Hawkeswood TJ, Knowles DG (1985) Observations on some jewel beetles (Buprestidae: Coleoptera) in Western Australia. *Victorian Naturalist* **102**, 205–206.

Hawkeswood TJ, Peterson M (1982) A review of larval host records for Australian jewel beetles (Coleoptera: Buprestidae). *Victorian Naturalist* **99**, 240–251.

Hawkeswood TJ, Sommung B, Sommung A (2018) First record of the jewel beetle, *Strigoptera bimaculata* (L., 1758) (Insecta: Coleoptera: Buprestidae) from Ubon Ratchatani Province, Thailand. *Calodema* **610**, 1–2.

Hawkeswood TJ, Turner JR (1994) A new species of the genus *Ethon* Laporte & Gory, with observations on its biology and host plants (Coleoptera: Buprestidae). *Giornale Italiano di Entomologia* **7**, 165–179.

Hawkeswood TJ, Turner JR (1997) Review of the biology, host plants, behaviour and parasites of the Australian buprestid beetle *Diadoxus erythrurus* (White) (Coleoptera: Buprestidae). *Mauritiana* **16**, S341–349.

Hawkeswood TJ, Turner JR (2009) *Castiarina lisaejessicae* sp. nov., a new jewel beetle (Coleoptera: Buprestidae) from New South Wales, Australia. (Coleoptera: Buprestidae). *Giornale Italiano di Entomologia* **12**, 199–203.

Heinrich B, Raven PH (1972) Energetics and pollination ecology. *Science, N.Y.* **176**, 597–602.

Hespenheide HA (2014) A reconsiderationof *Hylaeogena* Obenberger, 1923) (Coleoptera: Buprestidae), with descriptionof a new genus and new species from Mexico and Central America. *Coleopterists Bulletin* **68**, 21–30.

Hespenheide HA, Bellamy CL (2004) The first Antillean *Pachyschelus*, and a new *Leiopleura*, from Hispaniola (Coleoptera: Buprestidae). *Folia Heyrovskyana* **12**, 105–112.

Hill RS (1995) Conifer origin, evolution and diversification in the Southern Hemisphere. In *Ecology of Southern Conifers*. (Eds NJ Enright, RS Hill). pp. 10–29. Melbourne University Press, Carlton.

Hinz M, Klein A, Schmitz A, Schmitz H (2018) The impact of infrared radiation in flight control in the Australian 'fire beetle' *Merimna atrata*. *PLos One*. doi:10.1371/journal.pone.0192865

Hnatiuk RJ (1990) *Census of Australian Vascular Plants*. Australian Government Publishing Service, Canberra.

Hobler, FH (1925) A beautiful buprestid (Coleoptera). *The Queensland Naturalist* **5**, 42–43.

Hockey MJ, De Baar M (1988) Insects of the Queensland mangroves. Part 2. Coleoptera. *The Coleopterists Bulletin* **42**, 157–160.

Hołyński RB (1984) On Notogean and Oriental Mastogeniini LeC. Horn (Coleoptera: Buprestidae). *Polskie Pismo Entomologiczne* **54**, 105–114.

Hołyński RB (1988) Remarks on the general classification of Buprestidae Leach as applied to Maoraxiina Hol. *Folia Entomologica Hungarica* **49**, 49–54.

Hołyński RB (1992) A review of the genus *Paratrachys* Snd. (Coleoptera: Buprestidae). *Annals of the Upper Silesian Museum, Entomology* **3**, 115–136.

Hołyński RB (1993) A reassessment of the internal classification of the Buprestidae Leach (Coleoptera). *Crystal. Publications of the Natural Science Foundation at Göd. Series Zoologica* **1**, 1–42.

Hołyński RB (2008) A new species of New Guinea *Stigmodera* Esch. and remarks on *Pseudhyperantha* Snd. *Opole Scientific Society Nature Journal* **41**, 43–48.

Hołyński RB (2014a) A new species of the genus *Iridotaenia* Deyr. (Coleoptera: Buprestidae). *Genus* **25**, 387–414.

Hołyński RB (2014b) Description of three new subgenera and two new species of *Paracupta* Deyr. (Coleoptera: Buprestidae). *Genus* **25**, 403–414.

Hołyński RB (2014c) Supplementary remarks on the taxonomic structure of the genus *Chrysodema* C.G. (Coleoptera: Buprestidae). *Genus* **25**, 373–375.

Hołyński RB (2014d) New subgenus and three new species of the genus *Metataenia* Théry (Coleoptera: Buprestidae). *Genus* **25**, 391–402.

Hołyński RB (2022) Revision of the *Cyphogastra* Deyr. super genus (Coleoptera: Buprestidae) VI The Modita-, Obloquens-, Ventricosa- and Pistor- circles. *Procrustomachia* **7**, 1–38.

Hook AW, Evans HE (1991) Prey and parasites of *Cerceris fumipennis* (Hymenoptera: Sphecidae) from Central Texas, with description of the larva of *Dasymutilla scaevola* (Hymenoptera: Mutillidae). *Journal of the Kansas Entomological Society* **64**, 257–264.

Hope FW (1846) Descriptions of various new species of Buprestidae from Australia. *Transactions of the Entomological Society, London* **4**, 208–220.

House SM (1985) Breeding and Spatial Pattern in Dioecious Rainforest Trees. Thesis submitted for the degree of Doctor of Philosophy, Australian National University, Canberra.

Hunter RJ (2003) *World Heritage and Associative Values of the Central Eastern Rainforest Reserves of Australia*. New South Wales National Parks and Wildlife Service, Hurstville.

Hutchinson PM, Allsopp PG (2022) Two new species of *Castiarina* Gory and Laporte, 1838 (Coleoptera: Buprestidae: Stigmoderini). *Zootaxa* **5099**, 4.4. doi:10.11646/zootaxa.5099.4.4

Ivie MA, Miller RS (1984) Buprestidae (Coleoptera) of the Virgin Islands. *Florida Entomologist* **67**, 288–300. doi:10.2307/3493951

Jackman JA (1987) A new species of *Sambus* from Luzon Island, Philippines, with notes on species of other genera (Buprestidae: Coleoptera). *Coleopterist's Bulletin* **41**, 27–33.

Jendek E (2001) A comparative study of the abdomen in the family Buprestidae (Coleoptera). *Acta Musei Moraviae, Scientiae biologicae (Brno)* **86**, 1–41.

Jones J (1986) Evolution of the Fagaceae: the implications of foliar features. *Annals of the Missouri Botanical Garden* **73**, 228–275. doi:10.2307/2399112

Kang B, Kabir F, Bae E-J, Lee GS, Lee DW (2016) Damage report on a newly recorded Coleopteran pest, *Aphanisticus congener* (Coleoptera: Buprestidae) from turfgrass in Korea. *Weed and Turfgrass Science* **5**, 274–279. doi:10.5660/WTS.2016.5.4.274

Kerremans C (1896) Trachydes nouveaux. *Annales de la Societe Entomologique de Belgique* **xl**, 306–333.

Kerremans C (1902–1903) Coleoptera Serricornia Fam. Buprestidae. In *Genera Insectorum* (Ed. P Wytsman), 12, 1–338.

Kitchin DR (2009) Notes on the biology of *Merimna atrata* (Gory & Laporte) (Coleoptera: Buprestidae). *Australian Entomologist* **36**, 1–2.

Kitching RL, Boulter SL, Howlett BG, Goodall K (2007) Visitor assemblages at flowers in a tropical forest canopy. *Austral Ecology* **32**, 29–42. doi:10.1111/j.1442-9993.2007.01733.x

Kjemsmo K, Whitney HM, Scott-Samuel NE, Hall JR, Knowles H, *et al.* (2020) Iridescence as camouflage. *Current Biology* **30**, 551–555.

Knowles D (1984) Flying Jewels. *Geo* **5**, 46–57.

Kolibač J (2000) Classification and phylogeny of the Buprestoidea (Insecta: Coleoptera). *Acta Musei Moraviae, Scientiae biologicae* **85**, 113–184.

Kubán V, Majer K, Kolibač J (2001) Classification of the tribe Coraebini Bedel, 1921 (Coleoptera, Buprestidae, Agrilinae). *Acta Musei Moraviae. Scientiae Biologicae, Brno* **85**, 185–287.

Lacordaire JT (1857) *Histoire naturelle des insectes*. Librarie encyklopédique de Roret, Paris.

Ladiges PY, Udovic F, Nelson G (2003) Australian biogeographical connections and the phylogeny of large genera in the plant family Myrtaceae. *Journal of Biogeography* **30**, 989–998. doi:10.1046/j.1365-2699.2003.00881.x

Lang PJ (2020) *Buprestidae of South Australia*. at http://syzygium.xyzl/buprestidae/taxonomy.php. (Accessed March 2022).

Lang PJ, Stolarski AMP (2020) New host plants and distribution records in South Australia for *Microcastalia* Heller (Coleoptera: Buprestidae) with a summary of the taxonomic history of the genus. *Australian Entomologist* **47**, 195–211.

Lawrence JF (1988) Rhinorhipidae, a new beetle family from Australia, with comments on the phylogeny of the Elateriformia. *Invertebrate Taxonomy* **2**, 1–53. doi:10.1071/IT9880001

Lawrence JF, Britton EB (1994) *Australian Beetles*. Melbourne University Press, Melbourne.

Lawrence JF, Escalona HE (2019) Ommatidae Sharp and Muir, 1912. Chapter 2. In *Australian Beetles*, Vol. 2. (Eds A Ślipiński, JF Lawrence). CSIRO Publishing, Clayton South.

Lawrence JF, Lemann C (2019) Buprestidae Leach, 1815. Chapter 35. In *Australian Beetles*, Vol. 2. (Eds A Ślipiński, JF Lawrence). pp. 554–581. CSIRO Publishing, Clayton South.

Lawrence JF, Ślipiński A (2013) *Australian Beetles*. Vol. 1. CSIRO Publishing, Collingwood.

Lea AM (1895) Descriptions of new species of Australian Coleoptera. *Proceedings of the Linnean Society of New South Wales* **4**, 589–634.

Lea AM, Gray JT (1936) The food of Australian birds. An analysis of the stomach contents. *Pt iv. Emu* **35**, 251–280.

Levey B (1978a) A taxonomic revision of the genus *Prospheres* (Coleoptera: Buprestidae). *Australian Journal of Zoology* **26**, 713–726. doi:10.1071/ZO9780713

Levey B (1978b) A new tribe, Epistomentini, of Buprestidae (Coleoptera) with a redefinition of the tribe Chrysochroini. *Systematic Entomology* **3**, 153–158. doi:10.1111/j.1365-3113.1978.tb00111.x

Levey B (2012) A revision of the Australian species of the genus *Melobasis* Laporte & Gory 1837 (Coleoptera: Buprestidae). Part 1 (Introductory material, key to species groups and keys to species of the *thoracica*, *pusilla*, *formosa*, *propinqua gloriosa* species-groups). *Zootaxa* **3464**, 1–107.

Levey B (2018) A revision of Australian species of the genus *Melobasis* Laporte & Gory 1837 (Coleoptera: Buprestidae), Part 2 (Revision of the *nervosa* species group). *Zootaxa* **4528**, 001–079. doi:10.11646/zootaxa.4528.1.1

Levey B (2023) A revision of the Australian species of the genus *Melobasis* Laporte & Gory 1837 (Coleoptera: Buprestidae), Part 3 (Revision of the *azureipennis*, *cupricollis*, *iridicolor* and *melanura* species groups). *Zootaxa* **5302**, 001–100.

Levey B, Bellamy CL (2013) A taxonomic revision of *Neobuprestis* Kerremans (Coleoptera: Buprestidae) with a description of a new genus and two new species. *Zootaxa* **3681**, 225–240. doi:10.11646/zootaxa.3681.3.2

Macleay WJ (1872) Notes on a collection of insects from Gayndah. *Transactions of the Entomological Society of New South Wales* **2**, 239–318.

Macphail MK, Alley NF, Truswell EM, Sluiter IRK (2017) Early Tertiary vegetation evidence from spores and pollen. In *History of the Australian vegetation: Cretaceous to Recent*. (Ed. RS Hill). pp. 189–261. Cambridge University Press, Cambridge.

Macqueen J (1948) Notes of an interesting habit of the buprestid beetle *Stigmodera* (*Themognatha*) *fortnumi* (Hope). *Proceedings of the Entomological Society of Queensland* (1948), 1–3.

MacRae T (2006) Distributional and biological notes on North American Buprestidae (Coleoptera), with comments on variation in *Anthaxia* (*Haplanthaxia*) *cyanella* Gory and *A.*

(*H.*) *viridifrons* (Gory). *The Pan-Pacific Entomologist* **82**, 166–199.

MacRae T (2009) *Mastogenius quayllabambensis* MacRae, a new species from Ecuador (Coleoptera: Buprestidae, Haplostethini). *The Coleopterists Bulletin* **57**, 149–153. doi:10.1649/0010-065X(2003)057[0149:MGMANS]2.0.CO;2

Magallón S, Crane PR, Herendeen PS (1999) Phylogenetic pattern, diversity and diversification of eudicots. *Annals of the Missouri Botanical Garden* **86**, 297–372. doi:10.2307/2666180

Makinson RO (2018) Myrtle Rust reviewed: The impacts of the invasive plant pathogen *Austropuccinia psidii* on the Australian environment. Plant Biosecurity Cooperative Research Centre, Canberra.

Mallick S (2013) *Potential impacts of Climate Change on the fauna values of the Tasmanian Wilderness World Heritage Area.* Nature Conservation Report 13/2. Department of Primary Industries, Parks, Water and Environment.

Mast A, Thiele K (2007) The transfer of *Dryandra* R. Br. to *Banksia* L.f. (Proteaceae). *Australian Systematic Botany* **20**, 63–71. doi:10.1071/SB06016

Matthews EG (1972) A revision of the Scarabaeine dung beetles of Australia. 1. Tribe Onthophagini. *Australian Journal of Zoology, Supplementary Series.* Supplement 19, 1–330.

Matthews EG (1985) *Beetles of South Australia, Part 4.* South Australian Museum, Adelaide.

Matthews EG, Bouchard P (2008) *Tenebrionid Beetles of Australia: descriptions of tribes, keys to genera, catalogue of species.* Australian Biological Resources Study. Department of the Environment, Water, Heritage and the Arts, Canberra.

McKenna DD, Wild AL, Kanda K, Bellamy CL, Beutel RG, *et al.* (2015) The beetle tree of life reveals Coleoptera survived end Permian mass extinction to diversity to diversify during Cretaceous terrestrial revolution. *Systematic Entomology* **40**, 35–60.

Mecke R, Mille C, Engels W (2005) *Araucaria* beetles worldwide: evolution and host adaptations of a multi-genus phytophagous guild of disjunct Gondwana-derived biogeographic occurrence. *Pró Araucária Online* **1**, 1–18.

Mildenhall DC, Mortimer N, Bassett KN, Kennedy EM (2014) Oligocene paleogeography of New Zealand: maximum marine transgression. *New Zealand Journal of Geology and Geophysics* **57**, 107–109. doi:10.1080/00288306.2014.904387

Moore BP, Brown WV (1985) The buprestins: bitter principles of jewel beetles (Coleoptera: Buprestidae). *Journal of the Australian Entomological Society* **24**, 81–85.

Morgan FD (1966) The biology and behaviour of the beech buprestid, *Nascioides enysi* (Sharp) (Coleoptera: Buprestidae) with notes on its ecology and possibilities for control. *Transactions of the Royal Society of New Zealand (Zoology)* **7**, 159–170.

Mulder RH (1984) *Xyroscelis crocata*, an uncommon buprestid. *Circular of the Entomological Section, Royal Zoological Society of New South Wales* **34**, 4.

Nagalingum NS, Marshall CR, Quental TB, Rai HS, Little DP, *et al.* (2011) Recent synchronous radiation of a living fossil. *Science* **334**, 796–799. doi:10.1126/science.1209926

Nelson GH, Bellamy CL (1991) A revision and phylogenetic re-evaluation of the family Schizopodidae (Coleoptera, Buprestidae). *Journal of Natural History* **25**, 985–1026. doi:10.1080/00222939100770651

New TR (Ed.) (2007) *Beetle Conservation.* Springer, Dordrecht.

Nicholson AJ (1929) A new theory of mimicry in insects. *Australian Zoologist* **5**, 10–104.

Nikitin MI (1979) Buprestidae collected in the County of Cumberland. *Circular of the Entomology Section, Royal Zoological Society of New South Wales* **3**, 5–6.

Nylander U (2006) The New Guinean species of the genus *Castiarina* (Coleoptera: Buprestidae: Stigmoderini). *Lambillionea (Supplement 11)* **103**, 3–13.

Nylander U (2008) Review of the genera of *Calodema* and *Metaxymorpha* (Coleoptera: Buprestinae: Stigmoderini). *Folia Heyrovskyana Supplementum* **13**, 1–84.

Nylander U (2010) Notes concerning the genus *Metataenia* Théry, 1923 (Coleoptera, Buprestidae, Chrysochroina) from Papua New Guinea with description of a new species and designation of a lectotype. *Zootaxa* **2529**, 55–64. doi:10.11646/zootaxa.2529.1.3

Obenberger J (1923a) Une serie de noveaux genres de Buprestides (Coleoptera). *Acta Entomologica Musei Nationalis Pragae* **1**, 3–13.

Obenberger J (1923b) Studien über die Buprestiden. *Entomologische Blaetter fuer Biologie und Systematik der Kaefer* **19**, 114–125.

Obenberger J (1924a) Kritische studien über die Buprestiden (Col.). *Arch. Naturgesch* **3**, 1–171.

Obenberger J (1924b) Deuxieme serie de nouveaux genres de Buprestides. *Acta Entomologica Musei Nationalis Pragae* **1**, 13–44.

Obenberger J (1928) Opuscula Buprestologica I. Beiträge zur Kenntis der Buprestiden (Col.). *Arch. Naturgesch* **92**, 1–350.

Obenberger J (1930) Buprestidae II. In *Coleopterum Catalogus* (Ed. Schenkling S). pp. 215–568. W. Junk, Berlin.

Obenberger J (1941) Generis Bubastes Cast. et Gory revisio nova (Col. Bupr.). Nova revise rodu *Bubastes* Cast. et Gory (Col. Bupr.). *Acta Entomologica Musei Nationalis Pragae* **19**, 5–19.

Obenberger J (1955) Sur la morphologie du genre *Curis* Cast. et Gory avec les diagnoses des espéces nouvelles (Coleoptera, Buprestidae). *Acta Entomologica Musei Nationalis Pragae* **30**, 243–252.

Obenberger J (1958) Un genre nouveau de la sous-famille Polycestinae de l'Australie (Col. Buprestidae). *Acta Entomologica Musei Nationalis Pragae* **32**, 487–490.

Obenberger J (1959) Sur les espéces du genre *Agrilus* Curtis de l'Australie et Oceanie (Coleoptera, Buprestidae). *Acta Entomologica Musei Nationalis Pragae* **33**, 223–240.

Pan X, Chang H, Don R, Shih C (2011) The first fossil buprestids from the Middle Jurassic Jiulongshan Formation of China (Coleoptera: Buprestidae). *Zootaxa* **2745**, 53–62. doi:10.11646/zootaxa.2745.1.4

Pegg, G., Taylor, T., Entwistle, P., Guymer, G., Giblin, F. *et al.* (2017) Impact of *Austropuccinia psidii* (myrtle rust) on Myrtaceae-rich wet sclerophyll forests in south east Queensland. *PLos One* **12**, e0188058. doi:10.1371/journal.pone.0188058

Peterson M (1989) Taxonomic and correlated nomenclatural notes on *Stigmodera obesissima* Thomson and *Nascioides parryi* (Hope) (Coleoptera: Buprestidae). *Giornale Italiano di Entomologia* **4**, 205–212.

Peterson M (1991) A new southern Australian subgenus of *Temognatha* Solier (Coleoptera: Buprestidae: Buprestinae), with nomenclatural notes on Solier's generic name. *Records of the Western Australian Museum* **15**, 117–127.

Peterson M (1996) Aspects of female reproductive biology of two southwestern Australian *Temognatha* species (Coleoptera: Buprestidae). *Records of the Western Australian Museum* **18**, 203–2008.

Peterson M (2015) Revised taxonomy of *Temognatha duponti* (Boisduval) and *T. stevensii* (Gehin) (Coleoptera: Buprestidae: Stigmoderini) with definition of the *grandis* and *stevensii* species groups. *Journal of Insect Biodiversity* **3**, 1–25.

Peterson M (2018) Clarification of the type-locality of *Nascio chydaea* Olliff (Coleoptera: Buprestidae: Nascionini), with further notes on its ecology, distribution and relationships. *Journal of Insect Biodiversity* **008**, 035–042.

Peterson M, Bellamy CL (2000) Contribution to the taxonomy of Australian Buprestidae (Coleoptera). Part III. A new genus and species of buprestid from Queensland and comments about its placement. *Folia Heyrovskyana* **8**, 127–131.

Peterson M, Hawkeswood TJ (1980) Notes on the biology and distribution of two species of *Diadoxus* (Coleoptera: Buprestidae) in Western Australia. *The Western Australian Naturalist* **14**, 228–233.

Pineda C, Curletti G (2020) *Calodema autonkozlovi* nov. sp. (Coleoptera: Buprestidae): a new species of Stigmoderini from the Arfak Mountains, Indonesia. *Revista chilena de Entomologia* **46**, 75–79.

Pullen KR (1987) *Pseudotaenia waterhousei* V. d. Poll) (Coleoptera: Buprestidae) in New South Wales. *Australian Entomological Magazine* **14**, 23–28.

Queiroz JM (2002) distribution, survivorship and mortality sources in immature stages of the Neotropical leaf minor *Pachyschelus coeruleipennis* Kerremans (Coleoptera: Buprestidae). *Brazilian Journal of Biology* **62**, doi:10.1590/S1519-698422002000/00009

Ramasamy M (2019) A scientific note on occurrence and infestation of jewel beetle *Belionota prasina* (Coleoptera: Buprestidae) on Cashew (*Anacardium occidentale*). *National Academy of Science Letters* **42**, 91–94.

Ramírez-Barahona S, Sauquett H, Magallón S (2020) The delayed and geographically heterogeneous diversification of flowering plants. *Nature Ecology and Evolution* doi:10.1038/s41559-020-1241-3

Raven PH, Axelrod DI (1972) Plate tectonics and Australian paleobiogeography. *Science* **176**, 1379–1386.

Reyes-Gonzáles R, Toledo-Hernándes VH, Flores-Palacios A, Rös M, Bueno-Villegas J, *et al.* (2021) Testing three hypotheses of rarity among the Buprestidae species of a tropical dry forest. *Ecological Entomology* **2021**: 1–11.

Richards K, Spencer CP (2018) Exploitation of sapling *Banksia marginata* by *Cyrioides imperialis* (Fabricius 1801) (Coleoptera: Buprestidae) in Tasmania. *The Tasmanian Naturalist* **140**, 27–32.

Richards K, Spencer CP (2020) Jewels on fire! The Miena Jewel beetle *Castiarina insculpta* (Carter, 1934) (Coleoptera: Buprestidae), and the 2019 Great Pine Tier fire. *The Tasmania Naturalist* **142**, 35–40.

Richards K, Spencer CP (2021) First description of a male *Castiarina insculpta* (Carter, 1934) (Coleoptera: Buprestidae), a threatened Tasmanian jewel beetle. *Australian Entomologist* **48**, 289–292.

Robin L (1994) *Defending the Little Desert: The Rise of Ecological Consciousness in Australia*. Melbourne University Publishing Ltd., Melbourne. http://www.environmentandsociety.org/node/6811.

Ryczek S, Dettner K, Unverzagt C (2008) Synthesis of buprestins D, E, F, G and H: structural confirmation and biological testing of acyl glucoses of jewel beetles (Coleoptera: Buprestidae). *Bioorganic & Medicinal Chemistry* **17**, 1187–1192. doi:10.1016/j.bmc.2008.12.038

Saunders E (1871) *Catalogus Buprestidarum Synonymicus et Systematicus*. J. Janson, London.

Sauquet H, Cantrill DJ, Weston PH, Barker N, Mast A, *et al.* (2007) Bringing together the living and the dead: integrating extant and fossil biodiversity in evolutionary studies of Protcacac (Proteales). *Botany and Plant Biology 2007* Joint Congress, July7–11, Chicago.

Schlaepfer MA, Runge C, Sherman PW (2002) Ecological and evolutionary traps. *Trends in Ecology and Evolution* **17**, 474–480. doi:10.1016/S0169-5347(02)02580-6

Schmitz H, Schmitz A, Bleckmann H (2001) A new type of infrared organ in the Australian 'fire-beetle' *Merimna atrata* (Coleoptera: Buprestidae. *The Science of Nature* **87**, 542–545. doi:10.1007/s001140050775

Schneider ES, Schmitz H (2013) Bimodal innervation of the infrared organ of *Merimna atrata* (Coleoptera, Buprestidae) by thermo- and mechanosensory units. *Arthropod Structure and Development* **42**, 135–142. doi:10.1016/j.asd.2012.11.001

Schnepp KE, Ashman KL, Moore MR (2020) Report of an established population of *Belionota prasina* (Thunberg) (Coleoptera: Buprestidae: Buprestinae) in Florida, USA. *The Coleopterists Bulletin*, **74**, 124–126. doi:10.1649/0010-065X-74.1.124

Scobell L (n.d.) Mangrove dieback in the Gulf of Carpentaria. The Cairns and Far North Environment Centre, Cairns. <https://cafnec.org.au/wildlife-issues/mangroves-wetlands/mangrove-dieback-in-the-gulf-of-carpentaria/#:~:text=Over%207%2C000%20hectares%20of%20coastal%20mangroves%20have%20died,temperatures%20and%20unseasonally%20dry%20conditions%20in%20the%20region.> (Accessed 31 March 2023).

Simpson SR, Gross CL, Silberbauer LX (2005) Broom and honeybees in Australia: an alien liaison. *Plant Biology (Stuttgart)* **7**, 541–548. doi:10.1055/s-2005-865855

Ślipiński A, Lawrence JF (Eds.) (2019) *Australian Beetles*. Vol. 2. CSIRO Publishing, Clayton South.

Ślipiński SA, Leschen RAB, Lawrence JF (2011) Order Coleoptera Linnaeus, 1758. In *Animal Biodiversity: An Outline of Higher-Level Classification and Survey of Taxonomic Richness* (Ed. Z-Q Zhang). *Zootaxa* **3148**, 203–208.

Snow EL, Dhileepan K (2014) The jewel beetle *Hylaeogena jurecski*: a new biological control for cat's claw creeper (*Dolichandra unguis-cati*) in Queensland. In *Nineteenth Australasian Weeds Conference*. (Ed. M Baker) pp. 50–54. Tasmanian Weeds Society, Hobart.

Stilwell JA, Long JA (2011) *Frozen in Time: Prehistoric Life in Antarctica*. CSIRO Publishing, Collingwood.

Specht RL (1972) *The Vegetation of South Australia*. Government Printer, Adelaide.

Sundholm A, Catford A (1982) Jewel beetles: gems in a Western Australian Wilderness. *Habitat* **10**, 19–21.

Tarran M, Wilson PG, Hill RS (2016) Oldest record of *Metrosideros* (Myrtaceae): fossil flowers, fruits, and leaves from Australia. *American Journal of Botany* **103**, 754–768. doi:10.3732/ajb.1500469

Taylor TN, Taylor EL, Krings M (2009) *Paleobotany: The Biology and Evolution of Fossil Plants*. 2nd edn. Academic Press, Oxford.

Tepper JGO (1887). *Common Native Insects of South Australia: A Popular Guide to South Australian Entomology, Part 1, Coleoptera or Beetles*. E. S. Wigg and Sons, Adelaide.

Théry A (1923a) Note on the genus *Synechocera*, with description of a new species. *Proceedings of the Linnean Society of New South Wales* **48**, 517–518.

Théry A (1923b) Études sur les Buprestides. (Troisième partie). *Annals de la Société entomologoque de Belgique* **52**, 193–270.

Théry A (1928) A new buprestid from Australia. *Proceedings of the Linnean Society of New South Wales* **53**, 456–457.

Théry A (1929) Classification. In *A Checklist of the Australian Buprestidae*. (Eds. Carter HJ, Théry A). *Australian Zoologist* **5**, 266–275.

Théry A (1932) Contribution a l'etude des espéces du genre *Endelus* H. Deyr. (Coleoptera: Buprestidae). *Novitates Entomologicae* **2**, 1–23.

Théry A (1945) Descriptions de deux Buprestides des faunes australienne et mélanésienne (Coleoptera). *Bulletin de la Société Entomologique de France* **50**, 45–47.

Thornhill AH, Mcphail M (2012) Fossil myrtaceous pollen as evidence for the evolutionary history of Myrtaceae: a review of fossil *Myrtaceidites* species. *Review of Palaeobotany and Palynology* **176–177**, 1–23.

Thornhill AH, Simon YW, Külheim C, Crisp MD (2015) Interpreting the moder distribution of Myrtaceae using a dated molecular phylogeny. *Molecular Phylogenetics and Evolution* **93**, 29–43. doi:10.1016/j.ympev.2015.07.007

Tillyard RJ (1926) *The Insects of Australia and New Zealand*. Angus and Robertson, Sydney.

Tillyard RJ, Dunstan B (1923) Mesozoic insects of Queensland, Part 1 Introduction & Coleoptera. *Queensland Geological Survey* **273**, 1–75.

Toyama M (1986) The buprestid genus *Chalcophorella* Kerremans and its related genera (Coleoptera, Buprestidae). Entomological papers presented to Yoshihiko Kurosawa on the occasion of his retirement. Coleopterists' Association of Japan, Tokyo.

Toyama M (1987) The systematic position of some buprestid genera (Coleoptera, Buprestidae). *Elytra* **15**, 1–11.

Turner JR, Hawkeswood TJ (1994). Observations on the biology and host plants of *Dinocephalia cyanipennis* (Blackburn) (Coleoptera: Buprestidae) from New South Wales, Australia. *Giornale Italiano di Entomologia* **7**, 87–96.

Turner JR, Hawkeswood TJ (1996a) Taxonomy, biology, geographic distribution and conservation of the rare Australian beetle. *Stigmodera* (*Castiarina*) *armata* Thomson (Coleoptera: Buprestidae). *Giornale Italiano di Entomologia* **8**, 191–206.

Turner JR, Hawkeswood TJ (1996b) A note on the larval host plant and biology of the Australian jewel beetle *Astraeus crassus* Van de Poll (Coleoptera: Buprestidae). *Mauritiana* **16**, 75–79.

Volkovitsh M (2001) The comparative morphology of antennal structures in Buprestidae (Coleoptera), evolutionary trends, taxonomic and phylogenetic implications. Part 1. *Acta Musei Moraviae, Scientiae biologicae* **86**, 43–169.

Volkovitsh MG, Bílý S (2015) Larvae of Australian Buprestidae (Coleoptera). Part 5. Genera *Astraeus* and *Xyroscelis*, with notes on larval characters of Australian polycestine taxa. *Acta Entomologica Musei Nationalis Pragae* **55**, 173–202.

Volkovitsh MG, Bílý S, Hasenpusch J (2003) Larvae of Australian Buprestidae (Coleoptera). Part 2. Genus *Metaxymorpha*. *Folia Heyrovskyana* **11**, 203–216.

Volkovitsh MG, Hawkeswood TJ (1987) The larva of *Neocuris gracilis* Macleay (Coleoptera: Buprestidae). *Zoologischer Anzeiger* **219**, 274–282.

Volkovitsh MG, Hawkeswood TJ (1990) The larvae of *Agrilus australasiae* Laporte and Gory and *Ethon affine* Laporte and Gory (Insecta: Coleoptera: Buprestidae). *Spixiana* **13**, 43–59.

Volkovitsh MG, Hawkeswood TJ (1994) The larvae of *Melobasis* (*Melobasis*) *vertebralis* Carter (Coleoptera: Buprestidae). *Giornale Italiano di Entomologia* **7**, 11–27.

Volkovitsh MG, Hawkeswood TJ (1999) Larvae of *Prospheres aurantiopicta* (Laporte & Gory) with comments on the larval characteristics of Polycestoid taxa (Insecta, Coleoptera, Buprestidae). *Mauritiana (Atlenburg)* **2**, 295–314.

Walker K (2005) Jewel beetle (*Buprestis novemmaculata*). PaDIL – https://www.padil.gov.au/pests-and-diseases/pest/135607

Wang Lu-Yi, Franklin AM, Hugall AF, Medina I, Stuart-Fox D (2023) Disentangling thermal from alternative drivers of reflectance in jewel beetles: a macroecological study. *Global Ecology and Biogeography* **32**, 408–420. doi:10.1111/geb.13632

Webb GA, Simpson JA, Taylor EE (1988) Notes on the distribution and biology of *Theryaxia suttoni* Carter (Coleoptera: Buprestidae). *Australian Entomological Magazine* **14**, 98–99.

Westcott RL, Lavigne RJ (2019) Jewel beetles (Coleoptera: Buprestidae) as prey of robber flies (Diptera: Asilidae). *The Coleopterists Bulletin* **73**, 169–178. doi:10.1649/0010-065X-73.1.169

Westcott, RL, Romera-Napoles J, Jendek E (2016) Two observations of assassin bugs (Hemiptera: Reduviidae) feasting on adult jewel beetles (Coleoptera: Buprestidae) with notes on adults of other buprestid species and their predators. *The Coleopterists Bulletin* **70**, 384–386.

Whyte R, Anderson G (2017) *A Field Guide to Spiders of Australia*. CSIRO Publishing, Clayton South.

Williams GA (1977) A list of the Buprestidae (Coleoptera) collected from *Leptospermum flavescens* Sm. at East Minto, New South Wales. *Australian Entomological Magazine* **3**, 81–82.

Williams GA (1982) A note on Buprestidae (Coleoptera) observed at lights. *Australian Entomological Magazine* **8**, 81.

Williams GA (1983) Observations on the genus *Alcinous* Deyrolle (Coleoptera: Buprestidae). *Victorian Naturalist* **100**, 37–39.

Williams GA (1985) New larval food plants for some Australian Buprestidae and Cerambycidae (Coleoptera). *Australian Entomological Magazine* **12**, 41–46.

Williams GA (1987) A revision of the genus *Nascioides* Kerremans (Coleoptera: Buprestidae). *Invertebrate Taxonomy* **1**, 121–145. doi:10.1071/IT9870121

Williams GA (1993) *Hidden Rainforests*. New South Wales University Press, Kensington.

Williams GA (1995) Pollination ecology of lowland subtropical rainforests in New South Wales. Thesis submitted for the degree of Doctor of Philosophy, University of New South Wales, Sydney.

Williams GA (2002) *A Taxonomic and Biogeographic Review of the Invertebrates of the Central Eastern Rainforest Reserves of Australia (CERRA) World Heritage Area, and Adjacent Regions. Technical Reports of the Australia Museum*, 16.

Williams GA (2018) Insects associated with flowering of *Rhodomyrtus psidioides* (Myrtaceae): is this a Myrtle Rust (*Austropuccinia psidii*)–induced plant-pollinator interaction extinction event? *Cunninghamia* **18**, 023–027.

Williams GA (2020a) *The Invertebrate World of Australia's Subtropical Rainforests*. CSIRO Publishing, Clayton South.

Williams GA (2020b) Aspects of the reproductive ecology of a south-east Australian *Avicennia marina* mangrove community – flower visitors and potential pollinators. *Cunninghamia* **20**, 209–244.

Williams GA (2021) *The Flowering of Australia's Rainforests*. 2nd edn. CSIRO Publishing, Clayton South.

Williams GA, Adam P (1993) Ballistic pollen release in Australian members of the Moraceae. *Biotropica* **25**, 478–480. doi:10.2307/2388872

Williams GA, Adam P (1999) Pollen sculpture in subtropical rain forest plants: is wind pollination more common than previously suspected? *Biotropica* **31**, 520–524.

Williams GA, Adam P (1998) Pollen loads collected from large insects in Australian subtropical rainforests. *Proceedings of Linnean Society of New South Wales* **120**, 49–67.

Williams GA, Adam (2010) *The Flowering of Australia's Rainforests*. CSIRO Publishing, Clayton South.

Williams GA, Adam P (2019) A preliminary checklist of flower-visiting insects from *Syzygium floribundum*, *Syzygium smithii* and *Tristaniopsis laurina*: three members of the Myrtle Rust–vulnerable plant family Myrtaceae. *Cunninghamia* **19**, 057–074. doi:10.7751/cunninghamia.2019.19.005

Williams G, Bellamy CL (2002) New species and new records for the genus *Nascioides* Kerremans, 1903 (Insecta: Coleoptera: Buprestidae). *Zootaxa* **58**, 1–9. doi:10.11646/zootaxa.58.1.1

Williams GA, Watkins S (1985) A new species of *Nascioides* Kerremans (Coleoptera: Buprestidae). *Journal of the Australian Entomological Society* **24**, 255–259.

Williams GA, Watkins S (1986) A revision of the genus *Xyroscelis* Saunders (Coleoptera: Buprestidae). *Journal of the Australian Entomological Society* **25**, 295–299. doi:10.1111/j.1440-6055.1986.tb01118.x

Williams GA, Weir TA (1987) Four new species and new records of Australian Mastogeniinae (Coleoptera: Buprestidae). *Journal of the Australian Entomological Society* **26**, 153–159.

Williams GA, Weir TA (1988) Further new species of Australian Mastogeniinae (Coleoptera: Buprestidae). *Journal of the Australian Entomological Society* **27**, 179–181.

Williams GA, Weir TA (1992) Two new species of *Anthaxomorphus* Deyrolle (Coleoptera: Buprestidae: Trachyinae) from the Australasian region. *Journal of the Australian Entomological Society* **31**, 227–230.

Williams GA, Williams T (1983) A list of the Buprestidae (Coleoptera) of the Sydney basin, New South Wales, with adult food plant records and biological notes on food plant associations. *Australian Entomological Magazine* **9**, 81–93.

Willmer PG (2011) *Pollination and Floral Ecology*. Princeton University Press, Princeton.

Wilson PG, O'Brien MM, Gadek PA, Quinn CJ (2001) Myrtaceae revisited: a reassessment of infrafamilial groups. *American Journal of Botany* **88**, 2013–2025.

Yu Y, Ślipiński A, Shih C, Pang H, Don R (2013) A new fossil jewel beetle (Coleoptera: Buprestidae) from the Early Cretaceous of Inner Mongolia, China. *Zootaxa* **3637**, 355–360. doi:10.11646/zootaxa.3637.3.7

Zhu H-B, Gao Z-X, Han Z-C, Lu J, Chen J-D (2008) The jewel beetles (Buprestidae) intercepted on timbers from Papua New Guinea. *Chinese Bulletin of Entomology* **S41**, Q969.

Index

Page numbers in **bold** refer to figures, page numbers in *italics* refer to general discussions of genera in Chapter 2.

Aaaaba 5, 8, *87–8*, **138**

Aaaba see Aaaaba

Acmaeodera 1, 3

adaptations for pollen, nectar feeding 6

Agrilinae 2, **51–8**, 87–7, **138–41**, 169–70

Agrilus 1, 3, 4, 5, **51**, *88–9*, **139**, 147, 152, 153, 154, 157, 159, 160, 169

Alcinous see Aaaaba

amber, Buprestidae in 3

Anilara 4, 5, 8, **20**, *67*, **104**, 157, 159, 167, 168

Anthaxia 1, 67

Anthaxomorphus **52**, *89*

Anthaxoschema **20**, *68*

Aphanisticus 2, 6, **52**, *89*, 154

Araucariaceae 6, 7, 8, 62, 68, 94, 146, **174**

Araucariana 6, 8, **21**, *68*, **174**

Asamia 1

Astraeus 3, 4–5, 6, 7, 8, **15–16**, *59–60*, **98–101**, 147, 150, 157, 159, 160, 163, 171

Australorhipis **21**, *68–9*, **177**

Austrochalcophora 7, **18**, *63–4*

Austrophorella 7, **18**, *64*

Barakula 2, **21**, *69*

Belionota **21**, *69*, 144

Bubastes *69–70*, **104–5**, 157, 159, 160, 166

Buprestina *70*

Buprestinae 2, 3, 12, **20–34**, *67–82*, **104–8**, 166–9

Buprestis 1, **21**, *70*

Buprestodes 65

Buprestoidea 1, 2

Burnsiellus **22**, *70–1*, 160

Byrrhidae 2

Byrrhoidea 2

Callirhipidae 2

Callitris 4

Calodema 2, 3, 5, 9, 12, **34–5**, *82*, **109**, 144, 145, 184

Calotemognatha 12, **35**, *82–3*, 168

Castiarina 2, 3, 4, 5, 8, 9, 10, 12, **35–43**, *83–4*, **109–32**, **133**, 147, 148, 149, 151, 152, 153, 154, 155, 156, 157, 158, 160, 163, 168, 171, 172, 184

warning colour in 4

Chalcophorinae 2, **101–4**

Chalcophorotaenia 5, 7, **18**, *64*, **101–3**, 160, 163

chemical protection

buprestins 4

Chrysobothrinae 2

Chrysobothris 1, 3, 5, **22**, *71*, 154, 160, 167

Chrysochroinae 2, **18–19**, *63–7*, 163–6

Chrysodema **18**, *64–5*

Cisseis see Diphucrania

colour reflection 4

Conognatha **43**

conservation 12, 143, 157, 183–4

Curidini *67*

Curis see Selagis

Cyperaceae 4

Cyphogastra 7, **18**, *65*, **103**, 144

Cypriacis *70*, *see also Buprestis*

Cyria see Cyrioides

Cyrioides 4, 5, 6, 8, 10, **22**, *71–2*, **105**, 149, 154, 163, 167, 171

defence strategies 4–5

Diadoxus 4, 6, 8, **23**, *72–3*, **105**, 157, 160, 167

response to fire damaged trees 72–3

Dinocephalia 2, 4, 6, 7, 8, **52**, *89–90*, 157, 170, 171

Diphucrania 4, 5, 6, 8, 10, **52–3**, *90–1*, **139**, 149, 150, 152, 153, 157, 159, 160, 170, 171

Dystaxia **58**

Elateridae 2

Endelus 2, *91*

endemism 3, 9, 74, 157, 159, 162, 171

Ethon see Ethonion

Ethonion 5, 6, 8, 13, **54–5**, *91*, 149, 150, 171

Euchroma 1

Eucnemidae 2

Euphorbiaceae 5

Euryspilus **23**, *73*, **105**, 159

extra-continental associations 3–4

fossil buprestids 2–3

Australian fossils 3

fossil history 2–3

Galbellinae 1

Germarica 2, 4, 5, 7, 8, **55**, *60*, *91–2*, **139**, 150, 163

Gondwana 2, 3, 7–9, 59, 144, 151, 152

associations 3–4, 170

Habroloma 2, 5, 6, 8, 9, **55**, *92*, **140**, 154

Hedwidgiella **55**, *92–3*
Helferella 4, 5, **16**, *60–1*, 154, 163, 171, **174**
Hesperorhipis 2
Heteroceridae 2
Hiperantha **43**
Hypocisseis 4, 5, 10, **55–6**, *93*, **140**, 160
Hypostigmodera 83, 151, 157

Iridotaenia 7, **19**, *65*, 149

Julodimorpha 1, 2, 4, 6, **23**, *73–4*, 167
Julodinae 1
Julodis 1

Lampetis **19**, *65–6*, 144, 163
leaf mining 2
life histories 5–6
    gall formation 6
    as pollinators 11–13
Limnichidae 2

Maoraxia 2, 3, 6, 9, **23–4**, *74*, 154, 170
Mastogeninae 2, 60
Meliboeithon **56**, *93–4*, **140**, 169, **175**, 183
Melobasis 2, 3, 4, 5, 6, 8, 9, 10, 13, **24–7**, *74–5*, **105–6**, 146, 149, 152, 153, 154, 157, 159, 160, 162, 166, 171, **175**
Merimna 4, *75–6*
    response to fire damaged trees 76
Metataenia 66, *see also Paracupta*
Metaxymorpha 2, 3, 5, 8, 9, 10, 12, **44**, *84–5*, **132**, 144, 145, 184
Microcastalia **28**, *76*, 159
mimicry 4
myrmecophily, association with ants 6

Nascio 4, 5, **28**, *76–7*, 149, 153, 160, 166–7
Nascioides 3, 5, 6, 8, 9, **28–31**, *77–8*, **106–7**, 149, 151, 152, 153, 155, 157, 160, 170, 171, 172
Neobubastes *78*, **107**, 166
Neobuprestis 5, *78–9*, **108**, 151
Neocuris 2, 5, 8, 10, 12, 13, **31–2**, *79*, 155, 159, 163, 167
Neospades **56**, *94*, **140**, 160
Neotorresita see Pseudanilara
Nothofagaceae 6, 7, 8, 75, 77, 78, 170, **175**
Notobubastes **32**, *79*, 159
Notographus 5, **32**, *79–80*, 153, 168

Pachycisseis **56**, *94*, 157
Pachyschelus 1, **56**, *94*
Paracephala 6, 7, 8, **57**, *94–5*, **141**, 150
Paracupta 3, **19**, *66*, **103**, 153
parasitism 5
Paratrachys 1, 2, 3, 5, **17**, *61*, 154
pigmentation 4
plant associations 3, 5, 6–11, **173–81**
    conifers 6, 7, 11, 64, 183

cycads 6, 7, 11, 63, 163, **175**, **181**
    pollen carriage 13
    as pollinators of 11–13
    see also Araucariaceae, Gondwana, Nothofagaceae,
        Zamiaceae
Polybothris 1
Polycesta 3, **17**, *61*, 147
Polycestinae 2, 3, **16–18**, 59–63, **100–1**, 163
predation on 4–5
predators of 5
Prospheres 3, 5, 6, *61–2*, **101**, 153
    species, phylogenetic separation of 3–4
Prospherini 3
Psephenidae 2
Pseudanilara 5, 8, 9, **32**, *80*, 153, 154, 160, 168, 171
Pseudotaenia 2, 4, 7, **19**, *66–7*, **103–4**, 163, 166

Sambus **57**, *95*
Scaptelytra see Iridotaenia
Schizopodidae 1, 2
Schizopodinae 1
Selagis 2, 3, 4, 5, 6, 8, 12, **33**, *80–1*, **108**, 147, 157, 159, 160, 167
    South American associations 3
spiders 4, 5
    predators of Buprestidae 5
Stanwatkinsius 7, 8, 10, **57**, *95–6*, **141**, 170
Stenocera 1
Sternocerinae 2
Stigmodera 4, 5, 8, 12, **44–5**, *85–6*, **132–3**, 151, 156, 160, 168, 171, 184
Stigmoderinae 2, 3, 12, **34–50**, 82–7, 163
    as pollinators 5–6
    South American associations and taxa 3, **43**
Stigmoderini 2, 6
Strigoptera 3, 5, 7, **17**, *62–3*, 144
Synechocera 2, 4, 6, **57**, *96*, **141**, 153, 169, 171, **175**, 184

Temognatha 1, 2, 4, 5, 8, 10, 12, 13, **45–50**, *86–7*, **133–8**, 152, 156, 157, 159, 160, 162, 168–9, 171, 184
Themognatha see Temognatha
Theryaxia 5, 6, **34**, *81*
Throscidae 2
Torresita 4, 8, **34**, *81–2*, **108**, 155, 157, 160
Toxoscelus **57**, *96*, 144
Trachyinae 2
Trachys 2, 3, **58**, *96–7*, 153

warning colouration 4–5
wasps, parasites 5
wing reduction 1–2

Xyroscelis 3, 4, 6, **17–18**, *63*, **101**, 149, 163, 167, 184

Zamiaceae 6, 163, **175**, 184